Matthew Fontaine Maury

Manual of Geography

Matthew Fontaine Maury

Manual of Geography

ISBN/EAN: 9783741187353

Manufactured in Europe, USA, Canada, Australia, Japa

Cover: Foto ©Klaus-Uwe Gerhardt /pixelio.de

Manufactured and distributed by brebook publishing software
(www.brebook.com)

Matthew Fontaine Maury

Manual of Geography

MAURY'S GEOGRAPHICAL SERIES.

MANUAL

OF

GEOGRAPHY:

A

COMPLETE TREATISE ON MATHEMATICAL, CIVIL, AND PHYSICAL GEOGRAPHY.

BY M. F. MAURY, LL.D.,

Author of " Physical Geography of the Sea," etc.

UNIVERSITY PUBLISHING COMPANY,

NEW YORK AND BALTIMORE.

1878.

MAURY'S

GEOGRAPHICAL SERIES.

FIRST LESSONS IN GEOGRAPHY,

For Young Learners; in which the Author, in an imaginary voyage and journey, takes the pupil twice round the world, shows him various parts of it, and easily and pleasantly introduces him to the study of Geography.

THE WORLD WE LIVE IN:

An Intermediate Geography; in which the Author has sought to present the leading facts and principles of Geographical Science in a familiar and attractive manner, with constant reference to the maps, and with carefully adapted Questions, Exercises, and Map Studies.

MANUAL OF GEOGRAPHY:

A complete Treatise on Mathematical, Civil, and Physical Geography; presented in an attractive manner, with abundant helps and adaptations to awaken and sustain the interest of the pupil in intelligent study.

PHYSICAL GEOGRAPHY:

In which the Natural Features of the Earth, its Atmospherical Phenomena, and its Animal and Vegetable Life, are fully treated, with an attractiveness of style and freshness and interest of detail that charm the pupil and the general reader. Illustrated with numerous maps and engravings.

WALL MAPS:

With new and original features; furnishing invaluable aid in teaching Geography in classes, and comprising, I. The World. II. North America. III. The United States. IV. South America. V. Europe. VI. Asia. VII. Africa. VIII. Physical and Commercial Chart of the World.

Entered according to Act of Congress, in the year 1870, by

M. F. MAURY,

In the Office of the Librarian of Congress, at Washington.
·*· 333.

PREFACE.

As the matured fruit of the author's earnest and protracted labors this work is now sent forth, an humble contribution to the cause of geographical education.

The time seems fully to have arrived when geography demands an honorable place among her sister sciences. Every scholar of the present day is aware of the increased and increasing need of geographical text-books which, while within the intellectual grasp of young pupils, shall be fitted to expand the minds of those more advanced in their studies; and to redeem the most delightful of subjects from the bondage of dry statistics, on the one hand, and, on the other, from the drudgery of vague generalities.

In the preparation of this volume, as in that of its two predecessors, no pains have been spared to lead the young geographer by easy and gentle gradations to vantage-ground, from which he may overlook and survey nature for himself, and where the enchantments of the prospect will constrain him to pursue his geographical inquiries with zeal and enthusiasm.

"The study of physical phenomena," to borrow the words of Humboldt, "finds its noblest and richest reward in a knowledge of the chain of connection by which all natural forces are linked together and made mutually dependent on each other; and it is the perception of these relations that exalts our views and ennobles our enjoyments." While, therefore, the author has sought to reproduce in the pupil's mind the same vivid pictures of the various parts and places and objects of the globe which, as an eye-witness, he himself retains, he has constantly aimed at pointing out geographical laws, and at giving the learner glimpses into the terrestrial machinery, and frequent foretastes of the pleasures that await his after researches.

As regards the success which has attended the author's efforts to carry out his views, in these pages, the public must now judge.

The teacher and scholar, however, are alike requested to mark the following statements, which present some features of the Manual and furnish directions for its most effective use in the class-room.

(1.) *Map-Drawing* from memory is felt to be a *necessity* to all who would know the surface of the earth as illustrated by maps. The study of geography without a knowledge of the map is mere groping in the dark.

The Treatise on Map-Drawing here given is so simple in principle, and so easy in practice, that it cannot fail to commend itself to both teachers and pupils. Any one, however inexperienced, can use it, and pupils should be exercised in it from the beginning to the end of their course.

A uniform projection for maps is as desirable for schools and teachers as is a uniform system of weights and measures for business men—a consummation for which all nations are striving.

Universally, at sea, the Mercator Projection is used by navigators. That is the best for them. Likewise the Rectangular Tangential Projection is the best for the land. It is something new, and it is here presented for the first time, it is believed, in an American geography. The importance of it must not be measured by the space allotted to it. The attention of teachers is earnestly invited to it.

(2.) To avoid blurring the maps by dark shading and coloring in showing physical elevations, and to stimulate the pupil in getting the clearest ideas of the principal Mountain Chains and River Systems of the world, special maps have been prepared without regard to labor or expense.

These designs are entirely original, new, and unique, and have already been greatly admired and warmly commended by old and skillful teachers.

The various degrees of light and shade represent the elevations and depressions of the earth's surface; the darkest shades show the lowest lands and deepest valleys, and the lighter tints the higher lands. To make rivers perfectly distinguishable and traceable, they are marked in the darker shade by a white line, and in the lighter shades by a black line. Such representations of the earth's surface are indispensable in a school geography to the study of other maps.

(3.) To give greater elevation and zest to the study of the text, more than thirty Diagrams have been introduced explanatory of the earth's rotundity, of its revolution in its orbit, of the Stars of the Northern and Southern hemispheres, Isothermal Lines, the Great Lakes, the Trade Winds, the Monsoons, the Snow-line, the Barometer, Tides, Whirlwinds, the Size of Waves off the Cape of Good Hope, the differences of Time on different meridians, also some specimens of animal life in the sea.

Additional illustrations, drawn from life itself by the best artists of Europe and America, have been judiciously and lavishly inserted.

(4.) The Maps and the Map Studies are arranged to face each other.

(5.) The Map Studies are not mere questions on the map, but are among the most important pages of the book, on which the utmost care has been bestowed. To give them greater brightness and value, much pleasing matter, with occasional cuts and diagrams, have been thrown into them.

(6.) Pronunciation of difficult names has been generally given where they are first met with in the text; but there has been added at the close of the book a carefully compiled and judiciously selected Pronouncing Vocabulary.

(7.) Full Tables of Statistics have been appended to this work, but a large number of striking statistical data have been *interwoven with the text*.

The population and area of the different countries are furnished from the latest and best authorities.

(8.) The questions are *merely suggestive*, but to keep the pupil wide awake to all he has previously learned, it has been thought not unwise to ask occasional questions, especially in the Map Studies, which require him to examine other maps than the one just before him, and to draw upon the text.

(9.) The text has been broken up and marked by numbers and side-headings for convenience of reference.

(10.) A valuable Trade and Voyage Chart, exhibiting the great routes of commerce and their distances, the Ocean-Telegraph Cables, both finished and contemplated, the Currents of the Ocean and the Winds of the different Zones, with descriptive text, will be found near the close of the book.

(11.) Lastly, the *resumé* of the Most Recent Geographical Events and Discoveries up to the present time, by which, within the last two years, a new complexion has been put upon the geography of some portions of the earth, will be found specially interesting.

M. F. MAURY.

December, 1870.

CONTENTS.

MAPS. OROGRAPHIC VIEWS, ETC.

MAURY'S

MANUAL OF GEOGRAPHY.

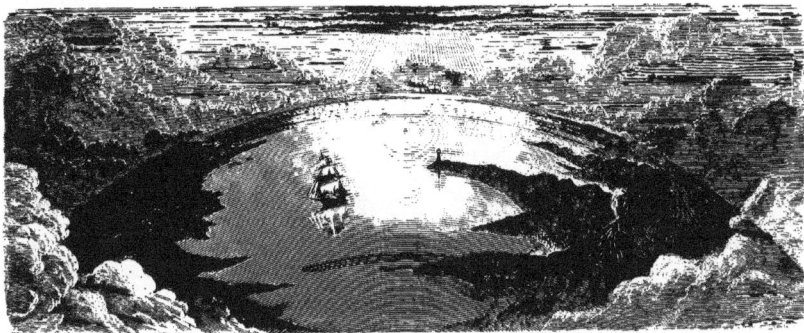

I. INTRODUCTORY.

LESSON I.

Definitions and Descriptions.

1. Geography is divided into mathematical, political, and natural or physical Geography.

2. Mathematical Geography treats of the shape and size of the earth, the determination of positions, and the measurements of distances and areas on its surface.

3. The Rotundity of the Earth.—Philosophers suspected that the earth was round, because, in watch-

SHIP SAILING FROM SHORE.

ing a ship departing from the shores of any country whatever, they had observed it to sink gradually below the horizon until, tips of the masts, which were the smallest, but the tallest parts of the ship, were all that could be seen. Early navigators thought the earth must be round, because, whenever they came in sight of land,

they first saw the tops of trees, or the needle-like summits of the mountains, while yet the huge dark masses of land beneath lay concealed from view.

In 1519, Ferdinand Magellan, a bold sailor of Portugal, confirmed these conjectures by actually sailing round the world.

Astronomers, finally, established them by remembering that the shadow cast by the earth, when it comes between the sun and the moon, so as to eclipse the moon, is as round as the shadow cast upon the wall by an orange; but, though the shadow of the earth is circular, and its form spherical, the exact shape and size of the earth had to be determined by laborious calculation.

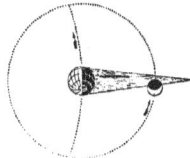

ECLIPSE OF THE MOON.

4. Definitions.—The *Circumference* of the earth is the distance around it.

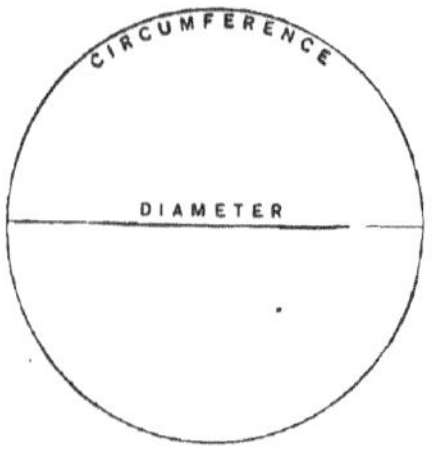

The *Diameter* of a circle is the distance through its centre, from one point on the circumference to the point opposite.

An *Arc* is any part of the circumference of a circle, as a rainbow.

A *Meridian* circle is one passing around the earth through the two poles.

5. *The Size and Shape of the Earth.*—From measurements which several Governments have caused to be made in various parts of the world as to the length of certain arcs of a meridian, it has been ascertained that the polar diameter of the earth is 26½ miles less than its equatorial diameter—which is 7,925¾ miles long—and consequently that the figure of the earth is that of an *oblate spheroid*.

An "oblate spheroid" is flattened at the poles, somewhat as an orange is at the stem, especially if it be slightly compressed between the finger and thumb.

6. *The Mean Circumference* of the earth is approximately and for convenience usually taken to be 25,000 miles.

7. *Area.*—Mathematical reckoning also tells us that the surface of a globe of such dimensions has an area of about 197,000,000 square miles.

Of this area, it is *estimated* by Geographers that about 145,000,000 square miles—or three-fourths of the whole area—are water, and the rest (52,000,000 square miles) *land*.

Questions.—What are the three principal branches of Geography?—What is Mathematical Geography?—Why did philosophers and navigators suspect the earth to be round?—How did Magellan confirm its rotundity?—By whom and how was this proof established?—What is the diameter of a circle?—What is an arc of a circle?—What is a Meridian?—Which is the longer, the Equatorial or the Polar diameter of the earth?—What then is the exact shape of the earth?—What is an oblate spheroid?—Can you calculate the mean circumference of the earth?—The diameter of a circle being multiplied by 3¹/₇—gives its circumference *nearly*—(exactly, if multiplied by 3.14159); can you tell by your own calculation what is the circumference of the earth at the equator?—How many square miles does the surface of the earth contain?—How much of this is supposed to be land, and how much water?

LESSON II.

Diurnal and Annual Motions of the Earth.

1. *Daily Rotation.*—It has been proved by observations on the stars, that the earth has a DIURNAL ROTATION from West to East, by which it makes a complete revolution on its axis once in every 24 hours. This period of time is called a day. As the circumference of the earth is 25,000 miles, a man standing on or near the equator will be moving toward the East at the rate of about one thousand miles an hour.

If an observer could watch our globe from the moon, and his eye first discern North and South America, these objects would, in a few hours, move out of sight; the Pacific Ocean would come into view instead; then the islands of Oceania, successively followed by Australia, Asia, Europe and Africa, the Atlantic Ocean and, finally, America again.

2. *Yearly Revolution.*—In addition to this axial rotation, the earth has its ANNUAL REVOLUTION round the sun, which it accomplishes once in every 365¼ days, or, more accurately, in 365 days, 5 hours, 48 minutes, and 50 seconds. This period makes a year, though we call 365 days a year, and correct for the fraction of a day (5 h., 48 m., and 50 s.) by adding another day every fourth year; this makes leap-year, which has 366 days.

But this allowance of ¼ of a day, or 6 hours, for every year is too much by 11 minutes and 10 seconds; to correct for this, so that the seasons may return *forever* at the same time of the year, it is necessary to count every 33d leap-year as a common year of 365 days, and, by skipping the one day, we cause the 21st of March and the 22d of September to be the days upon which the equinoxes must always occur; and so, for all time, mid-summer is made to fall in July, and mid-winter in January.

3. *Earth's Orbit.*—The path of the earth in its annual revolution around the sun is called its ORBIT. This orbit is not a circle, but an oblong called an ellipse, and the distance of the sun from it, and consequently from us, is about 91,500,000 miles. The diameter of a circle is a little less than one-third of its circumference: by doubling 91,500,000 of miles you get the diameter of the circle that the earth annually describes in its orbit around the sun. Now multiply this by 3¼, then make the calculation, and you find that in our annual journey around the sun, we are travelling at the rate of more than a thousand miles a minute.

The motion of the earth in its annual revolution causes the seasons, and its diurnal rotation on its axis causes day and night.

4. *Day and Night.*—The sun is always shining on one half of the earth, and then it is day; while the other half, being turned away from him, is in its own shadow—that makes it dark, and then it is night. When the sun is directly overhead, it is said to be *vertical*. The line in which an apple would fall from the top of a tree to the ground is called a *perpendicular*.

The phenomena of the solstices, the equinoxes, and the seasons are also easy of explanation. You will learn about them in the next lesson.

Questions.—How many motions has the earth ?—What is its diurnal rotation ?—How long does it take to make a complete rotation on its axis ?— Is this rotation from East to West, or in the contrary direction ?—Which motion causes day and night ?—What is its annual motion ?—How long does it take the earth to make a complete revolution around the sun ?—How do we correct for the fraction of a day ?—How much is it ?—What is the orbit of the earth ?—What is the shape of the orbit ? (If the teacher have a globe in the school-room, he should use it to illustrate the motions which cause day and night, summer and winter, etc.) How far is the earth from the sun ? Can you tell how far the earth, in its orbit around the sun, travels in a minute ? When is the sun *vertical ?*—What do you understand by a *perpendicular line ?*

LESSON III.

The Axis of the Earth and the Seasons.

1. Axis.—The earth, in its diurnal rotation, turns upon its polar or shorter diameter, as the spinning-top turns upon its own axis. This polar diameter is what is meant by the axis of the earth.

A PLANE, in mathematical geography, means an out-stretching level, like an immense floor or a perfectly flat meadow of boundless extent.

2. The Inclination of the Earth's Axis to its Orbit.—The axis of the earth is inclined to the plane of its orbit, as the axis of a leaning top is inclined to the floor.

The leaning top spins round on its axis, and travels round some point on the floor ; the floor is the plane in which the top revolves. In like manner, the earth wheels round on its axis in diurnal rotation in the plane of its orbit, and travels round the sun in that plane in annual revolution. Now there is this difference between the earth and the top : the top inclines more and more as its spinning slacks, but the earth never slacks its rate, and the inclination of its axis to this plane is always the same. It inclines from the perpendicular at the constant angle of 23° 28' ; and our north pole constantly points to the north star.

This may not be so always. If all things continue as they now are 12,000 years longer, a bright star called *Vega*, in the constellation known as Lyra, will be our polar star. There are seven stars called the "seven point-ers" or "the dipper," two of which point directly toward the north star. The first clear night look toward the north, and see if you can find "the dipper," and tell the north star by it. As the dipper never sets in this country, it may be seen any clear night.

It is by virtue of this simple contrivance of the Divine Architect in inclining the earth's axis, that the year is

THE INCLINATION OF THE EARTH'S AXIS.

divided into seasons. I told you that the cause of the seasons was as simple in explanation as day and night, and here you have it all fully exemplified by the leaning top as it spins about the floor.

DIPPER AND NORTH STAR.

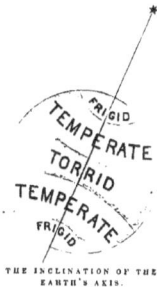

If the earth's axis were perpendicular to the plane of its orbit, as the axis of the sleeping top is to the plane of the floor, the days and nights would be of equal length all the year ; neither would there be any change of seasons.

Refer to the diagram, Lesson IV., or to a globe, and you will understand how at one season of the year the north pole is toward the sun, and at another season the south pole ; and that, therefore, the sun in his annual round *appears* to travel from north to south—being high up in the heavens at noon in summer, and low down in winter—whereas the sun is standing, and the earth is moving under him.

3. The Tropics and the Solstices.—It is owing to this *apparent* motion of the sun from one tropic to the other and back, that he is so high in the heavens at noon in summer, and so low in winter. When he reaches the highest point in summer and the lowest in winter, he appears to *stand still,* for he gets neither higher nor lower at *noon* for several days. One of these "*stand-still*" places is called the *summer solstice,* and the other the *winter solstice (sun-stand).*

Thus you see how the revolution of the earth around the sun, combined with the inclination of its axis, causes the seasons.

At the summer solstice the sun at noon is directly overhead to all places in lat. 23° 28' north. Here he appears to stop, to *turn back* and begin to go south again. This turning place is on the TROPIC OF CANCER.

Tropic is from a Greek word which signifies *to turn.*

The TROPIC OF CANCER is a circle drawn around the earth parallel to the equator, and at every point exactly 23° 28' distant from it.

In like manner, when the sun reaches the winter solstice, it is vertical at noon to all places in lat. 23° 28' south ; and a circle drawn here around the earth and parallel to the equator, is called the TROPIC OF CAPRICORN. These two circles are 46° 56' (twice 23° 28') from each other.

The sun is never vertical to any place north of the Tropic of Cancer nor to any place south of the Tropic of Capricorn.

4. The Zones or Belts of the Earth.—The belt of the earth between these two parallels of latitude is called the TORRID ZONE. It embraces an area of about 78,000,000 sqr. miles, or two-fifths of the entire surface of the earth. *These are the Inter-tropical regions.*

The sun is vertical twice a year to all places within these regions, and there is no cold weather; it is summer all the year round and the people do not, as a rule, even build chimneys to their houses.

At the same distance from each pole, viz.: 23° 28', there are two other circles drawn parallel to the equator. The one about the north pole is the ARCTIC CIRCLE, and the one about the south pole is the ANTARCTIC CIRCLE.

The area embraced between each of these circles and its nearest pole measures 8,000,000 sqr. miles.

The space that lies between the Arctic Circle and the north pole is the NORTH FRIGID ZONE. In it the summers are short and cold, and the winters long, dreary, and severe, and as you approach the pole the days become longer and longer, till you get where they have but one day and one night during the whole year, each being six months long.

The same is the case with the SOUTH FRIGID ZONE, which lies between the Antarctic Circle and the south pole.

These two zones together contain an area of upward of 16,000,000 sqr. miles, most of which has never been trod by human foot, or seen by the eye of man. Consequently we do not know whether these unexplored regions contain most land or most water. But we do know that when the sun shines at one pole, it is night at the other.

Questions.—What is the axis of the Earth?—What is a Plane?—What is the inclination of the earth to the plane of its orbit?—Suppose the axis of the earth were perpendicular to the plane of its orbit, what effect would that have upon the seasons, and the length of day and night?—How many *solstices* are there?—When do they occur?—What do you mean by a solstice?—What are the *Tropics?*—In what hemisphere is the tropic of Cancer?—How far is each Tropic from the Equator?—How far from each other?—What are the regions called that lie between the tropics?—How many sqr. miles does the torrid zone contain?—What portion is this of the entire surface of the earth?—Is the climate of the Inter-tropical regions all winter or all summer?—Describe the Arctic and the Antarctic Circles. Where are the Frigid Zones?—What is their area?—Describe the climates there and the length of the days.

LESSON IV.

The Equinoxes.

1. The Vernal Equinox.—Owing to the inclination of the earth's axis to the plane of its orbit, the sun, as you have been told already, *appears* to move up and down the heavens from the tropic of Capricorn to the tropic of Cancer, and back, once a year. (See Hemispheres, Less. V., and point out the two tropics.)

Though this motion is only *apparent*, yet for the convenience of explanation we will consider it as real.

In consequence of this inclination, the sun, in passing from the tropic of Capricorn to the tropic of Cancer, crosses the Equator on its way to the North. This happens on the 21st of March every year; on that day the sun sets at the south pole and rises at the north pole.

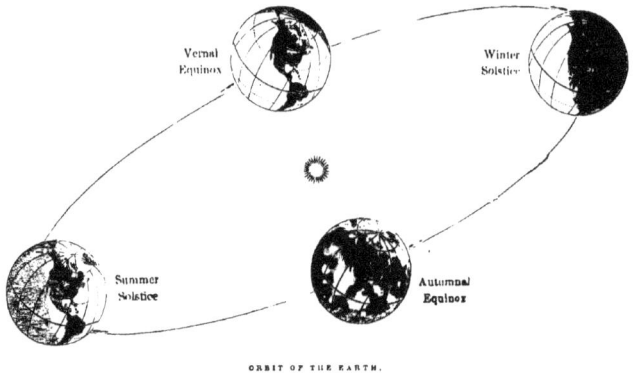

ORBIT OF THE EARTH.

At all other places on that day it rises and sets at six o'clock, consequently the day and night are then equal: this is the VERNAL EQUINOX.

2. The Autumnal Equinox.—Six months afterward —on the 22d Sept.—as the sun returns from the tropic of Cancer to the tropic of Capricorn, it again crosses the Equator, when it sets at the north, and rises at the south pole: day and night are again equal, and this is called the AUTUMNAL EQUINOX.

3. *Seasons.*—Thus the year is divided into seasons, and the seasons on the two sides of the Equator are opposite; that is, when it is winter with us in the NORTHERN HEMISPHERE, it is summer with the people on the other side of the Equator, in the SOUTHERN HEMISPHERE.

HEMISPHERE means *half* sphere, and we can divide the earth into Northern and Southern halves, as well as into Eastern and Western.

As the earth's orbit is an Ellipse, the earth is not always at the same distance from the sun. It is about 3,000,000 of miles nearer the sun in winter than in summer; but, in winter, the Northern half of the planet leans farthest away from the sun, and also receives his rays less vertically than in summer—hence it is colder.

4. *The Temperate Zones.*—The region embraced between the tropic of Cancer and the Arctic Circle is called the NORTH TEMPERATE ZONE. That between the tropic of Capricorn and the Antarctic Circle is the SOUTH TEMPERATE ZONE.

The North Temperate Zone is the one in which we live. All parts of the United States, except the northern portion of Alaska, lie within it; and in it, as

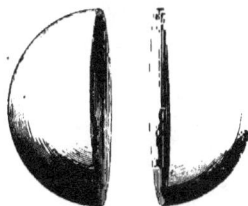

THE HEMISPHERES.

you know, we have summer and winter with the pleasing diversity of seasons.

102,000,000 square miles, or a little more than half the earth's surface, is contained in these two zones.

5. *The Frigid Zones.*—When the sun, in his apparent motion, goes south of the Equator after the 22d of September, as has been indicated, darkness settles down upon the North Frigid Zone, and night reigns for six months with uninterrupted gloom.

At the time of the Vernal Equinox, when the season for his return draws near, the cheerless inhabitants of these icy lands anxiously look for him, and are said to climb mountains to catch a glimpse of his earliest beams.

2

SUN AT MIDNIGHT IN FRIGID ZONE

When he rises upon them in the spring, it is also for six months, during which time they have no night.

These circumstances of day and night, occur in reversed order in the South Frigid Zone.

Questions.—When do the Equinoxes occur?—In what month is the *Vernal* and in what the *Autumnal* Equinox?—How long are the days then? Describe the season in the Southern hemisphere when it is winter with us. Where are the Temperate zones?—How much of the earth's surface do they contain?—In what zone do we live, and in which hemisphere?—Describe the day and night in the Frigid zones.

LESSON V.
Study of the Hemispherical Maps.

It is impossible for a scholar to make satisfactory progress in Geography without constant reference to maps. Next to visiting all parts of the earth and seeing the objects themselves, the best thing is closely to inspect pictures or drawings which represent them. In beginning our map-studies, for convenience, we divide our globe into two parts, called the Western and Eastern Hemispheres. When Columbus sailed on his voyage of discovery, as you have already learned, he sailed to the West, and, consequently, the new country he found was called the Western World.

Eastern means turned toward the point where the *sun rises.*

Western, turned toward the point where the *sun sets.*

WESTERN HEMISPHERE.

By what meridian circle have we here divided the earth into hemispheres? *Ans.* By that 20° west of Greenwich (near London, Eng ; see p. 108).—What great land-masses do you find in the western hemisphere?—What in the eastern hemisphere?—What great body of land lies partly in each hemisphere? —*The islands of the Pacific Ocean from the American coasts to the 95th meridian of east longitude, form Oceania.*—Which is the largest island in Oceania?

In which hemisphere is Greenland?—In which is Spitzbergen?—Victoria Land?—The Unexplored Regions?

In what direction does North America lie from Europe ?—From Africa ? —In what direction is China from the United States?—In what is Australia? What grand land-masses lie wholly north of the Equator?—What are

divided by that geographical line into two parts?—What immense island lies wholly south of the Equator?

What is the most northerly cape of Europe?—Which is the most southerly cape of South America?

What grand divisions of the earth are intersected by the Tropic of Cancer?— What, by the Tropic of Capricorn?—What, by the Arctic Circle?—Where does this circle touch Europe?—Are there any large bodies of land intersected by the Antarctic Circle?—Through what parts of the world, both land and water divisions, does the 60th meridian of west longitude pass?—Through and near what parts does the 150th meridian pass?—Through what parts does the 90th meridian of east longitude pass?—Through what does the meridian of Green-

EASTERN HEMISPHERE.

wich pass? That of Washington?—Through what parts of the world does the meridian of Teneriffe, one of the Canary islands, pass?

How do the Pacific and Atlantic Oceans compare as to shape? *Ans.* The Pacific is long and wide; the Atlantic is so very *narrow*, in proportion to its length, that geographers often call it the *Atlantic Canal.*—On what oceans would you sail in going from America, in a southeastwardly course, to Australia, thence to California?—What oceans would you sail on in a voyage from New York to California, by way of Cape Horn?—From New York to China, by way of the Cape of Good Hope?—From Cape Horn to the Cape of Good Hope?

Name the political divisions here given of North America.—Name those of South America—Of Europe—Of Asia—Of Oceania—Of Africa—What large islands lie in the Pacific Ocean west of the United States?—What islands lie west of the coast of Africa?—What, northeast of the United States?—What, southeast of the United States?—What large island lies east of Africa?—What large island north of Australia?—Where is New Zealand?—Borneo?—Java?—Sumatra?—What islands together form a great Empire in the Pacific Ocean? *Ans.* The Japan islands.—What large islands in the Arctic Ocean are begirt with ice? *Ans.* Nova Zembla and Spitzbergen.

What are the most southerly capes of Africa?—What, the most easterly?—What, the most southerly cape of India?—Where is Kerguelen's Land? Falkland islands?—Isle of France?—Where is the Strait of Sunda?

LESSON VI.
Latitude and Longitude.

1. The Equator is a circle passing from West to East round the earth midway between the poles. It divides the earth into two equal parts, one called the Northern, and the other the Southern hemisphere.

2. Any circle that divides the globe into two equal parts, is called a **Great Circle.** All meridians are great circles.

3. A **small circle** is any circle that divides a sphere into two *unequal* parts. The tropics of Cancer and Capricorn and all parallels of latitude are small circles.

The position of places on the earth's surface is designated by their latitude and longitude. Parallels are parallel to the equator.

The ancients supposed the earth was longer from East to West than from North to South.

4. The Latitude of a place is its distance, expressed in degrees (°--″) from the Equator. If on the North side, the place is in North Latitude; if on the South side, it is in South Latitude.

The pole is at 90° of latitude. No place can have more than 90° of latitude, because no place can be farther from the Equator than the pole. Those regions of the earth lying within the tropics and near the Equator, are said to be *low latitudes*. *High latitudes* are those near the Arctic and Antarctic circles and the poles.

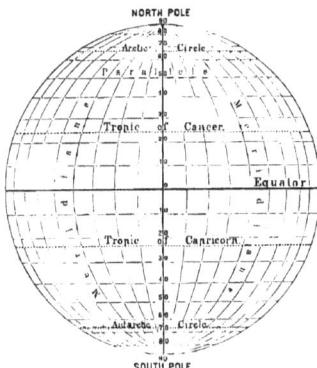

PARALLELS AND MERIDIANS.

5. Parallels of latitude are circles that pass round the earth, parallel to the Equator. The lines that pass from left to right across every map are parallels of latitude.

Point out a parallel of latitude on the map, Less. V.

6. A Prime Meridian is any meridian from which a nation may choose to reckon longitude.

7. The Longitude of a place is its distance, likewise in degrees, from the *Prime Meridian*. If the place be East of the prime meridian, it is in East Longitude, and if on the other side, it is in West Longitude.

Places on a prime meridian have no longitude whatever.

8. A Meridian is a great circle that crosses the Equator at right angles and passes through the poles. Those lines that run from North to South on maps are Meridians of Longitude.

Point out meridians of longitude on the map, Less. V.

So you see that the lines on the map that run from one side to the other are Parallels of Latitude, those that run from top to bottom are Meridians of Longitude.

And you see, moreover, by looking at the map, that all meridians cross each other at both the north and the south pole, each of which is 90° from the Equator. The distance between the poles is therefore 180°.

9. Degrees of Longitude.—As all the meridians cross at the poles, and diverge or spread out thence till they reach the Equator, the distance between any two varies with the latitude. Therefore a degree of longitude is greater at the Equator than it is anywhere else.

TABLE
English miles to a degree of Longitude for every 5th degree of Latitude from the Equator to the Poles.

Lat.	Miles.	Lat.	Miles.	Lat.	Miles.
0°	69.19	30°	59.9	60	34.5
5°	68.7	35	56.7	65°	29.8
10	67.9	40°	52.3	70	23.6
15°	66.9	45°	48.9	75°	17.8
20°	65.0	50°	44.4	80°	11.9
25°	62.8	55°	39.3	85°	5.35

At 90°, or the Pole, there is no such thing as Longitude.

The sun, in its *apparent* motion, passes over 15° of longitude every hour, whether those degrees are taken on the Equator or near the poles.

As all parts of the globe move together from West to East, toward the sun, it is plain that a short degree of longitude near the pole, since it moves slowly, will occupy as much time, in passing under the sun, as a long degree will require for its passage.

10. Sea Miles.—It is usual to reckon the length of a degree of longitude or latitude in miles of 60 to a degree at the Equator. These are called sea, geographical, or nautical miles. All nations use them at sea, because their use facilitates to the navigator his calculations. Nautical miles are also called *knots*, because the marks on the log line, by which the speed of ships at sea is measured, consists of knots spliced into the line. A *nautical* or *geographical* mile is longer than a *common* mile, which we call a statute or an *English* mile; for while there are (nearly) 60¼ statute miles to a degree of longitude at the Equator, there are only 60 nautical miles.

Distances by sea and the length of telegraphic cables are usually expressed in nautical miles.

11. A Marine League is three nautical miles. The jurisdiction of every country that fronts on the sea extends out to the distance of a marine league from the shore. Vessels of nations at war cannot join battle, or commit any act of hostility, within that distance of the shores of a neutral power. This distance was fixed upon, by the common consent of nations, and under the idea that no cannon could· ever send a ball farther than one marine league from the shore ; and that every nation has the exclusive right of jurisdiction over as much sea as guns on her shores could command. But the rifle cannon, and improved ordnance of the present day, can send their shot much farther than three miles. All beyond this marine league is what is·called the *high seas*, which, like the air, is free to all the world, and to which no nation,,however powerful, has, an exclusive right, any more than the farmer has·to the common highway which passes through his land.

12. Reckoning Longitude.—The· Prime Meridian, we have seen, is any meridian from which a nation may choose to reckon longitude The Germans reckon longitude from the meridian of Ferro, the most westerly of the Canary group ; the French from the meridian of the Paris Observatory; the English from Greenwich (*Grēn'-itch*); the Spaniards from Cadiz, etc. We have two Prime Meridians—one for the land, the other for the water.

We reckon longitude *at sea* from the meridian of Greenwich ; and *on land* we reckon the longitude of all places in the United States from the meridian of Washington. Washington is 77° west of the meridian of Greenwich.

Most of the charts used by mariners at sea are constructed at the Hydrographical Office in London, and from the meridian of Greenwich. For the convenience of navigation, therefore, we use the meridian of Greenwich on all our charts.

Here is the convenience and advantage of using latitude and longitude and maps and charts. If you were to say you had met a person on the railroad, no one could tell the place of meeting ; but if you were to say you met at the *crossing* of a certain other road, every one would know the exact spot.

So it is with the geographer, he designates the position of places on the earth by the crossing of parallels of latitude and meridians of longitude with each other. Thus, if you were told that a' ship was spoken at sea, in Lat. 40° north, for example, nobody could point out the place ; but if you were told she was spoken lat. 40° north

and long. 30° west, you would understand that she was spoken at the very spot where the parallel of 40° north crosses the meridian of 30° west ; and you would then see that the ship, when spoken, was in the Atlantic Ocean, and near the Azores.

Without this mode of marking positions on the chart, navigators would never find their way across the seas.

13. The Mariner's Compass helps them to this.

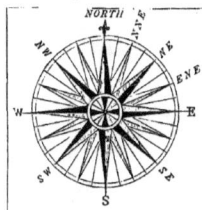

MARINER'S COMPASS.

The card upon which the courses are written is attached to a magnetic needle below. This needle points toward the north and south, and if it varies from the true north—as it generally does—the amount of deviation can always be determined, either at sea or on land, by astronomical observation.

You observe that the edge of this card toward the top of the page, is marked " North," that toward the right hand of it is marked " East," the bottom is marked " South," and the left hand " West."

These four points—N., E., S. and W.—are called the *Cardinal Points*.

Seamen divide the compass into 32 points, and each point into halves and quarters. Where nicety and accuracy are required, the compass card is still farther subdivided into degrees. But for the ordinary purposes of geography the points are sufficient. They are marked on the compass, and to tell them in order, beginning at the north and going around to the right ; as, N., NNE., NE., ENE., E., etc., etc., and so on all the way round to the north again, is called *boxing* the compass.

Questions.—Turn to the map (Less. V.) and point out the Equator ; the Tropics; the Torrid Zone; the Temperate Zones; the Frigid Zones. Point out Parallels of Latitude; what are they? Meridians of Longitude; what are they?—What and where is the Equator?—Into what two grand divisions does it divide the earth?—What is a small circle?—Are all small circles parallels of latitude?—What small circles are parallels of latitude?—What latitude has the pole?—How do you designate the geographical position of places?—How do you reckon latitude?—What is a Prime Meridian?—How do you reckon longitude?

Are degrees of longitude of an invariable length?—What is the length in nautical miles of a degree of longitude at the Equator?—Ditto in statute miles?—Over how many degrees of longitude does the sun pass every hour?—What do mariners mean by "*knots*," when they say their ship is going so many knots an hour?—Do they mean statute miles or sea miles?—What is a marine league?—How far out to sea does the jurisdiction of a nation extend from her shores?—Why was it limited to this distance?

From what meridian do the French reckon longitude?—From what the English?—From what meridian do we reckon longitude?—Why do we use the meridian of Greenwich for charts?—There is an island on the Equator in long. 120° E.; can you find it on the map, and tell what island it is?—(Less. V.)—There are some islands and a sea between the parallels of 40° N. and the Arctic Circle, and in long. 180°—What are they?—Can you *box* the compass?—What are the four cardinal points?

1. Continent.	5. Cape.	9. Peninsula.	13. Bay.	17. Lake.
2. Island.	6. Promontory.	10. Shore.	14. Strait.	18. River.
3. Hill.	7. Mountain.	11. Ocean.	15. Sound.	19. Delta.
4. Isthmus.	8. Volcano.	12. Sea.	16. Channel.	20. Archipelago.
		21. Table Land.		

LESSON VII.

Natural Geography.

1. Natural Geography, in the sense here meant, treats of the surface of the earth in its natural aspects.

2. Land and Water.—The Almighty, at the creation, made of the earth two grand divisions—land and water. For convenience the geographer has subdivided the water into sheets of various forms and sizes, which he has named Oceans, Seas, Bays, Gulfs, Harbors, Lakes, and Rivers; and the land into Continents, Islands, Peninsulas, Capes, Mountains, Valleys, Deserts and Plains.

These are all usually called natural divisions.

3. Flora and Fauna.—The surface of the earth is clothed with vegetation and animated with living creatures: these are its *Flora* and its *Fauna.*

4. Mines.—Its crust also is stored with coal and marble, copper, iron, the ores of metals, gold and precious stones: these are called minerals.

It belongs to natural geography to treat of all these, and to show the industries connected with them, as well as how the fauna of a country depend upon its flora.

This dependence, you see, is obvious; for every animal, whether it be insect, bird, or beast, requires food that is suitable to it.

Therefore, a country that produces no grass can have no flocks; and lions cannot subsist unless they have flocks to prey upon. Hence lions are to be found only in grass countries. In like manner, bees, hummingbirds, and flowers go together ; and if you take away the flowers, the birds and the insects disappear.

These relations and dependencies are very interesting, and it is the business of the geographer to study them all.

Questions. —What are the grand natural divisions of the earth ?—How is the water divided ?—How the land ?—Can you point out on the map (Lesson V.) some of each of these natural divisions ?—What do you mean by the flora of a country ?—What by the fauna ?—What by the minerals ?—Can you cite cases to illustrate how the fauna of a country depend upon its flora ?

LESSON VIII.

Definitions in Natural Geography.

1. Mercator's Chart. —You remember that the earth is a sphere, and you saw that the maps, Less. V., attempt to represent on a plane the surface of the earth. This makes the countries near the edge of the map appear, as compared with those near its centre, out of proportion, and it throws places out of their true relative position both as to course and distance. It was almost impossible for navigators to find by such a chart their true course and distance from port to port. This being the case, Mercator, a native of Antwerp in Belgium, invented the chart that goes by his name.

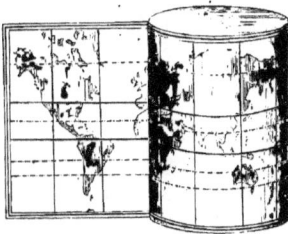

MERCATOR'S CHART.

It, too, distorts the surface it represents, as every chart must do that attempts to represent on a plane the surface of a sphere ; but it distorts in such a manner as to make all places on it preserve their true course from each other. This also makes it easy to take their true distance apart.

The charts that navigators use to sail by at sea are all constructed upon the Mercator principle.

2. Divisions of Water. —You observe by looking at this chart that all parts of the sea are really connected with each other ; and though the seas are all one sheet of water, yet they have been divided into five grand

divisions called Oceans, viz., the Pacific, the Indian, the Atlantic, the Arctic, and the Southern Oceans.

3. Old and New Worlds. —The land comprises two grand divisions : the " OLD WORLD" and the " NEW WORLD." But geographers for convenience have divided the two grand masses of land into four natural sections, and called them Continents, viz., North America, South America, Europe and Asia, and Africa.

4. Bays and Seas. -Some parts of the ocean are called Seas, as the Mediterranean Sea, the Caribbean Sea, the North Sea ; others, Bays : as Hudson Bay, the Bay of Biscay, the Bay of Bengal ; others, Gulfs : as the Gulf of Mexico, the Gulf of California, the Gulf of Finland. The definition that is applied to a sea is equally applicable to a bay and gulf, viz., it is a sheet of water or arm of the sea partly surrounded by land.

5. A Strait, Pass. or Passage. is a narrow channel that connects two larger sheets of water, as the Strait of Gibraltar, the Strait of Babelmandeb, Behring's Strait, the Florida Pass, the Mona Passage, the Windward Passage, etc., etc.

6. A Harbor is a sheltered arm of an ocean, sea, bay, or gulf, where ships may anchor and ride in safety. It is generally named after the town or city which is situated upon it, as Boston Harbor, Annapolis Harbor, the Harbor of Rio de Janeiro, the Harbor of Liverpool.

7. A Lake is a large inland sheet of water, either with or without a river running out of it, as Lake Superior, the Great Salt Lake, the Lake of Geneva. Sometimes the level of a lake is far above the level of the sea, and sometimes below it.

Questions. —Why can you not represent without distortion the surface of the earth on a map ?—Why do navigators prefer charts on the Mercator principle to charts of any other construction ?—In which respect, then, does the Mercator chart truly represent places ?—What and how many are the grand divisions of the water ?—Of the land ?—What definitions would you give to a sea, bay, or a gulf ?—What to a strait, pass, and passage ?— What to a harbor ?—To a lake ?

LESSON IX.
The Land.

1. A Continent is a large body of land, large enough to contain Empires and States, and so extended that you cannot sail round it.

2. An Island is land that is surrounded by water : as the Islands of Newfoundland, Great Britain, Nantucket. There are innumerable Islands.

3. A Peninsula (*pene*, almost ; *insula*, island), is

land that is *almost* surrounded by water. Yucatan is a Peninsula, Nova Scotia is a Peninsula, Portugal and Spain together form a Peninsula.

4. *A Cape, Point, or Headland,* is the extreme end of any land that juts out into the sea—as Cape Hatteras, Cape Henry, Cape Cod, Cape Horn, etc.

The Lizard, in England, is a POINT, which is among mariners one of the most famous landmarks in the world.

5. *Rivers* are natural gutters and drains, for carrying the water back to the sea after it has performed the manifold offices that Supreme wisdom and goodness have assigned to it.

6. *Mountains* are the watersheds which the same Almighty Builder has constructed for turning the rains off into their appropriate drains and channels, and so making the land inhabitable.

It is interesting to know how high a mountain may be, as it is to know how long, deep, and wide, a river is. The height of mountains is always reckoned perpendicularly from the level of the sea.

NOTE.—The sea is taken as the standard plane, because the elevation of the land is, in the process of ages, likely to change. We know that all of the Mississippi valley, and even the tops of what are now very high mountains, were once at the bottom of the sea. But the mean level of the sea is always the same; and by measuring the height of the land above it, we have already discovered that in some parts of the world the land is now gradually rising up, while in other parts it is sinking down.

7. *The Height of Mountains* is generally ascertained by the difference between the weight of the atmosphere on the sea-shore, and its weight on the mountain top. This difference is called the difference of *barometric* pressure.

8. *The Atmosphere has weight* as water has, but it is not so heavy and you do not perceive it, for the same reason that a swimmer does not feel the weight of water above him when he dives—nevertheless the atmosphere presses with the force of nearly 15 lbs. upon every square inch of your body.

NOTE.—The "Barometer" is an instrument for measuring the pressure of the atmosphere. The mean pressure or weight of the atmosphere at the surface of the ocean is almost 15 lbs. to the square inch; and under that pressure the quicksilver in the tube of the Barometer stands at the height of 30 inches. Now as you carry the Barometer up above the sea, as in a balloon or up the side of a mountain, the quicksilver will fall in the tube about one-tenth of an inch for every 90 feet of perpendicular ascent.

9. *Deserts* are wide, flat wastes of land, covered with sand, generally destitute of vegetation and water.

They are perilous to the traveller; but, as we shall see, when we come to study physical geography, indispensable parts of the earth's machinery.

Natural Geography (sometimes called Physical Geography) treats also of the Sea, Earth, and Air, in all the aspects in which they present themselves to an observer of nature.

In this sense Physical Geography is one of the most interesting, instructive, and profitable studies in the whole course of education.

A separate and special treatise will be devoted to it. In this work it is treated in its more limited sense as *natural* geography, which takes cognizance simply of the land and water as they are presented to the eye by nature and represented on the map.

Questions.—What is a Continent—an Island—a Peninsula—a Cape? In what light does the physical geographer regard rivers—in what, mountains? How and from what plane is the height of mountains measured?—Why do you measure it from the plane of the sea instead of from the surrounding plains?—How much does the atmosphere usually press upon a square inch at the level of the sea?—Does it press more or less than this upon the top of a high mountain?—What is the name of the instrument that measures this pressure?—To what does physical geography chiefly relate?—How is physical geography treated here?

Map Exercises.—Point out on the map the islands and peninsulas that are named in the lesson; also the deserts and the mountains that are on the map.—Tell—judging by the map and the eye—which is the largest ocean; the smallest continent; the largest island; the largest range of mountains.—In what light should rivers and mountains be regarded by the physical geographer?—Can you name any useful purpose that they serve?—Why is the sea-level adopted as the standard from which the height of mountains is measured?—What is the mean pressure of the atmosphere when measured at the level of the sea by the Barometer?—What is the most common way of measuring the height of mountains?

LESSON X.

Political Geography and the different kinds of Religion in the World.

1. *Political Geography* treats of the inhabitants of the earth; of their manners and customs, their industrial pursuits, their religion, and their forms of government.

It treats also of the divisions which have been made on its surface by the various nations, as Empires, Kingdoms, and States—such for example as Russia, France, the United States, etc.; and of the features which have been, by man's agency, impressed upon it—as canals, railroads, and other memorials of his handiwork—cities and towns, farms, mines, dwellings, edifices, manufacturing and other establishments, with which his industry and energies have embellished the landscape.

2. *Religions.*—All people have some kind of religion. Those who believe that Jesus Christ is the Son of God and who worship the Creator, as we are taught to do in the Bible, are called Christians. Those who do not believe the Bible, and who worship idols, or any of the objects of nature—as the sun, birds, beasts, or fishes, or who worship spirits and imaginary beings, are called heathens or pagans.

All the nations of America without exception, and all the nations of Europe except Turkey, profess the Christian religion.

3. Christendom.—The Christian countries, that is, those lands where God is professedly worshipped according to the Bible, constitute what is called Christendom.

Of the 1350 millions in the world—its estimated population—Christendom contains about 360 millions.

Many individuals among this vast multitude are unbelievers and atheists, so that we may assume that not one-fourth of all the people in the world acknowledge that Jesus Christ is the Son of God.

The remaining three-fourths—990,000,000—are beyond the pale of Christianity. They are heathens, who have various creeds and many forms of worship.

4. Judaism.—The Jews believe in the Old Testament, but not in the New. They worship God, but hold that Christ was a man, and that the Saviour is yet to come. It is estimated that there are 4,000,000 Jews in the world, most of whom live in Europe and America.

5. Islamism.—The Mahometans believe that there is one God, and that Mahomet is his prophet. Mahomet wrote the Koran about 600 years after Christ and the Koran is the Bible of his followers.

The Turks, the Arabs, the Persians, and the inhabitants generally of the *dry* countries of Asia and Africa, are followers of Mahomet. They are estimated to number about 60,000,000.

6. Buddhism.—Buddha was the founder of this religion. His followers do not believe in any God. Some of the peoples who follow him have no word in their language for Deity or immortality. They believe there is no being superior to man; and the object of their religion is to show the way to Nirvana, where man is annihilated. According to the Buddhist, annihilation would be the summit of bliss.

It is estimated that one-third of the entire human family profess this religion. The majority of the people of Burmah, many in India, China, Japan, Ceylon, Siam, etc., etc., are Buddhists.

7. Brahminism.—Its followers, next to the Buddhists, are the most numerous. They too are dwellers

in Asia. The Brahmins hold sway in India. They have deities, some of which, according to their doctrines, have previously been incarnate, sometimes as men with many hands or as beasts with many heads.

Juggernaut is one of their most famous idols. He is mounted on a car, and his worshippers on certain occasions fall down before it, that it may roll over and crush them to death as his worshippers haul him along. The Brahmins number about 150,000,000 souls.

8. The Guebres (Güe'berz).—These are the followers of Zoroaster. The sun is the object of adoration with them. They are the fire-worshippers of Persia and India, and they are to the heathen nations what the Jews are to the Christian nations—a people without a country, but who in a manner still preserve their nationality.

It should not be inferred from this sketch that there are no Christians except in Europe and America. There are many. The English and other European powers have established colonies and settlements in various parts of Asia and Africa, and among the islands, all of which acknowledge the Christian religion. Moreover, the missionaries of Europe and America have made many converts to Christianity in most heathen lands. But there are among the 960,000,000 souls who are supposed to inhabit Asia and Africa, no nations except one who regard the Bible or acknowledge the religion taught by it ; and that one is Liberia, which consists of a few thousand negroes, most of whom are emancipated slaves of the Southern States, sent thence to Liberia since 1823, by the Colonization Society of the United States.

Out of every one hundred souls in the world only twenty-six belong to the States of Christendom. This includes the infidels, atheists, heathens, and unbelievers of all sorts that dwell in Christian lands.

Questions.—Of what does Political Geography treat ?—Point out on the map (pp. 20, 21) some of the political divisions of the earth.—Name some of its political features.—Which two of the continents are inhabited chiefly by Christians ?—Name the chief Christian nations.—Who are the Mohammedans ?—What is the chief Mohammedan nation ?—How many inhabitants is the world supposed to contain ?—How many of them are Christians ?—How many pagans or heathens ?—What do they worship?—What is the creed of the Buddhists ?—What peoples profess this religion ?—What is the number of them ?—Point out on the map the parts of the world inhabited by them ?—Who are the Mohammedans?—Who wrote the Koran ?—When was it written ?—Who and what are the Brahmins ?—Tell about Juggernaut.—Who are the Fire-worshippers ?—Have they any country of their own ?—Name the founders of each one of these sects.—How many Jews are there supposed to be in the world ?—Have they any country of their own ?—In what parts of the world do most of them dwell ?—Which, the people that inhabit Christian countries or heathen lands, are the most numerous?—Are all the inhabitants of Christendom Christians ?—Are there any Christians in heathen lands ?

3

LESSON XI.

Governments.

1. Savages.—In some parts of the world the people are ignorant of the proper distinction between right and wrong. You, who have never lived among savages, can have no idea how brutal and ignorant some of them are. When the Georgian Islands in the Pacific were first discovered, the natives did not even know the use of fire.

2. Government and its End.—But all people, whether savage or civilized, who live together, either as families, tribes, or nations, require government of some sort; otherwise the strong will oppress the weak.

3. Kinds of Government.—There are among civilized nations various modes of accomplishing this end. All of them differ more or less in detail: but civilized governments of the present day may be reduced to two kinds—the Republican and the Monarchial.

4. Republican Government, as that of the United States, is based upon the doctrine that all good government rests upon the consent of the governed.

5. Congress.—Every two years we have a new Congress, and the President is chosen for four years.

Congress and the President make the laws, but they can only make laws in relation to such matters as the States have, in the Constitution, empowered them to legislate upon.

6. Monarchies.—In England, the crown is hereditary, and the government is based upon the assumption of the law that the "King can do no wrong." When there are abuses in administration the country holds the ministers, and not their sovereign, responsible for the proper administration of public affairs.

7. Senate and House of Representatives.—With us Congress consists of the Senate and House of Representatives. Two Senators are chosen by each State to serve six years at a time, and the Representatives are chosen by the people to serve two years, each State sending representatives in proportion to her population.

8. Parliament.—In England, Parliament is the Congress. It consists of the House of Lords and the House of Commons. The members of the latter are chosen by the people for seven years, and in the former the members consist of certain of the nobility—the Peers of the realm.

9. The English Government is what is called a *Limited Monarchy ;* so are all the monarchies of Europe, except Russia and Turkey.

10. Absolute Monarchies.—Those are *Absolute Monarchies* where the will of the Sovereign is the supreme law of the land.

There are Empires, as Russia ; but there is no difference between an Empire and a Kingdom, except in name.

As a rule, an Empire is supposed to be larger than a Kingdom.

11. The Nations of Europe, including Great Britain, are three Empires, thirteen Kingdoms, and two Republics; besides the petty Republics of San Marino and Andorra, there are twenty-four Duchies, petty Principalities, and free States.

12. The Nations of America.—These are seventeen republics, including San Domingo, and one empire.

13. The Nations of Asia.—These are two empires and eight kingdoms that are recognized by us as belonging to the family of nations. But besides these there are an almost infinite number of tribes and petty powers.

14. The Nations of Africa.—Africa comprises one republic, several provinces, and a number of so-called kingdoms and tribes that do not attain to the dignity of nations.

Questions.—Are any people too ignorant to have some sort of government?—What is the main object of government?—What are the two principal kinds of governments?—Under what sort of government do you live?—Tell about Congress.—Under what sort of government do the English live?—How many independent nations are there in Europe?—Point out the empires.—How many nations in America?—Point them out on the map.—How many in Asia?—Look at the map, and name them.

LESSON XII.

The Industrial Pursuits of Man, and the Geographical Distribution of Labor.

1. Human Industries.—In most countries the chief industrial pursuits are agricultural. But agriculture is by no means the only branch of industry among nations and people. With some, mining ; with others, manufacturing ; and with others, seafaring, is the chief branch of industry.

2. Chief Branches of Industry. — The following are considered the chief branches of industry in a geographical point of view :

AGRICULTURE, which includes tilling the earth and the raising of flocks and herds.

SEAFARING, which includes fishing and navigation.

MINING, the raising of ores and minerals from the bowels of the earth, and refining them.

MANUFACTURES, the working up of all sorts of raw material, so as to bring them into more useful shapes.

COMMERCE includes buying and selling, and the exchanging of the products of one country for those of another.

3. Why Industries vary.—To understand the influences which cause people to turn their attention to this or that branch of industry, is, practically, one of the most important and useful of the many highly instructive branches of geography.

In Louisiana, for example, the cultivation of the sugar-cane is an important branch of industry.

In New England the cutting and gathering of ice from the ponds in winter, the putting of it on board ships, and the sending of it off to different parts of the world for a market, is an important industry.

There are no ice-ponds in Louisiana and no cane-fields in New England, simply because the laws of nature, as expressed by climate, forbid. For this reason the rural industries of countries differ.

4. Industries depend on Geographical Conditions.—All industries depend upon geographical circumstances and upon the natural resources, situation, and exigencies of the country.

As an illustration: Great Britain abounds in coal; and that you may understand what an important part coal plays in her industries, please to remember that while it took one hundred thousand men, Herodotus says, twenty years to build the Great Pyramid of Egypt, it takes Great Britain only nine days to raise coal enough to make a pile just as grand. In a single year she has raised not less than one hundred and twelve millions of tons of coal. Now, remember, that a greater part of this coal is used by her people in driving machinery and in manufacturing, and then reflect that there is in one pound of coal power enough to do as much work in one hour as ten able-bodied men can do in a day, and you will not be surprised to hear that the Island of Great Britain, though not half so large as California, contains fifty times the number of people, who sell annually to other nations more than one thousand millions of dollars' worth of manufactured goods.

Questions.—What, in a geographical point of view, are the five chief branches of human industry?—Are these industries distributed about the world in obedience to any of the laws of nature?—Name one of the principal branches of industry in Louisiana.—In Maine.—Why cannot the sugar-cane be cultivated in Maine?—Why will it not grow there?—How many tons of coal are raised in Great Britain annually?

LESSON XIII.

About the Influences which Regulate the Geographical Distribution of Labor.

1. How Nations become Great.—We have seen what a great nation the English people have, by conforming their industries to their geographical surroundings, established on an island that is not as large as any one of the largest States of our Union.

2. Climate.—This, more than any other single cause, influences the geographical distribution of human labor. Nature has prescribed for every shrub, for every tree, and for every animal, except man alone, its geographical range.

3. The "Geographical Range" of a plant is the extent of the earth's surface within which that plant will thrive in the open air. Each kind of plant, as well as each kind of animal, has its special geographical range. Thus the geographical range of the chinchona tree, of cochineal, of the india-rubber tree, and of the pineapple, is confined to the Torrid Zone.

The division of the earth into zones is an artifice of man: these are not natural divisions, like hill and vale, land and water, and nature does not recognize them.

4. List, etc.—The following list contains the names, first, of the staple vegetable productions that have their geographical ranges confined to the zones, and, second, of those that overlap the dividing lines, and are found in more than one zone.

1. TORRID ZONE.—India-rubber, gutta-percha, spices, bamboo, cacao, cochineal, coffee, the plantain, bread-fruit, cherimoya and mangosteen—the most delightful of all fruits—sarsaparilla and cinchona, sago, opium, dye-woods, mahogany, pineapples, limes, mangoes, palm-oil, and the aloe.

2. TEMPERATE ZONE.—Hemp, flax, buckwheat, naval stores (tar, pitch, and turpentine are called naval stores), maple sugar, madder, mulberries, currants, etc.

3. FRIGID ZONE.—The vegetation of this zone is very poor, and none of the great agricultural staples, except perhaps barley, will grow there.

4. PRODUCTS THAT OVERLAP THE ZONES.—The staple vegetable productions that are found both in the Torrid and Temperate Zones, are—sugar, tea, rice, cotton, coffee, corn, wheat, indigo, tobacco, oats, peas, beans, barley, rye, pomegranates, melons, apples, pears, peaches, plums, oranges, cherries, lemons, strawberries, figs, potatoes, beets, turnips, carrots, parsnips, the vine, pumpkins, onions, almonds, and a great variety of other nuts, small fruits, flowers, and vegetables.

It is chiefly in obedience to climate that all labor, except mining, is distributed over the earth.

5. Division of Animals.—We divide animals into two classes, the *Graminivorous* and the *Carnivorous*.

Graminivorous means feeding on grass; Carnivorous, feeding on flesh. The former, as the horse and the monkey, live upon grass and other vegetable food. The latter, as the dog and the lion, live upon flesh. The range of *carnivorous* animals is, in like manner, limited by the geographical range of the graminivorous animals which serve them for prey, and the range of these last by the range of the plants upon which they feed.

6. Man alone without Geographical Range.—All animals but man have a limited range. Man can live and move everywhere, in the Hot and in the Frigid Zone, in the desert and in the swamp, in the depth of the mine and on the loftiest mountain-top.

Glaisher, the English aeronaut, ascended more than five miles over London; but the highest dwelling of man is not over three miles above the sea.

Questions.—What is it that chiefly influences the distribution of human labor on the earth?—Give an example of how labor depends upon climate.—What is meant by the geographical range of a plant or animal?—What animal has the widest geographical range?—Name some of the plants and animals that have their geographical range confined to a particular zone.—Are the five zones natural or artificial divisions?—Name some of the plants that overlap these artificial divisions.—Which of the five chief branches of industry is least influenced by climate?—Why?—What are the graminivorous and what the carnivorous animals?—The range of man?

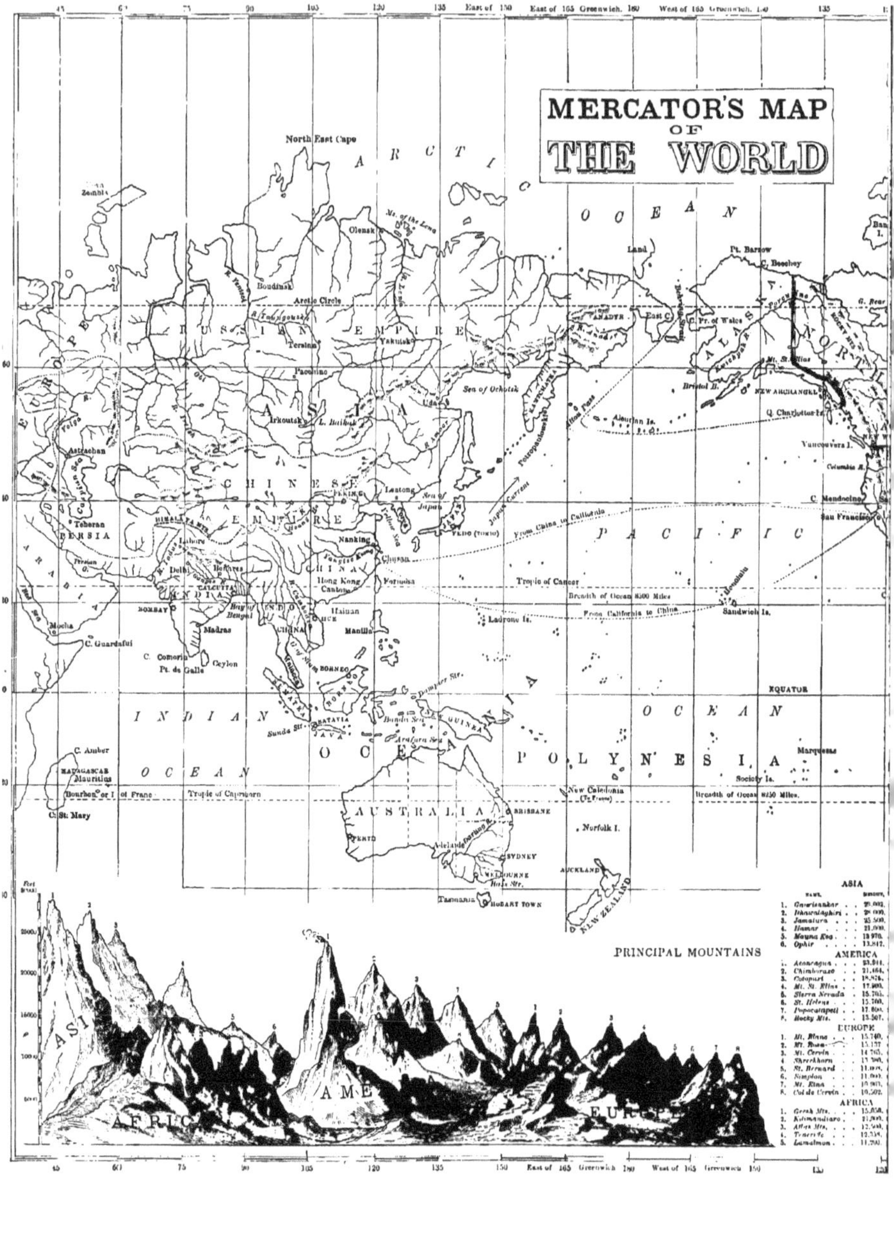

MERCATOR'S MAP
OF
THE WORLD

A R C T I C   O C E A N

North East Cape
Land
Pt. Barrow
Behring
C. Bear

Arctic Circle
Olenak
RUSSIAN EMPIRE
Terminal
Yakutsk
Paoutino
Arkoutsk
L. Baikal
Sea of Okhotsk

NEW ARCHANGEL
Q. Charlotte Is.
Vancouvers I.
Columbia R.
C. Mendocino
San Francisco

Astrachan
CHINESE EMPIRE
PEKIN
Leaoung
Sea of Japan
Nanking
CHINA
Hong Kong
Canton
Formosa
Haiman
Macao

Teheran
PERSIA
Persian Gulf
Delhi
Lahore
Benares
INDIA
BOMBAY
CALCUTTA
Bay of Bengal
Madras
Ceylon
Pt. de Galle

P A C I F I C   O C E A N
Tropic of Cancer
Ladrone Is.
Breadth of Ocean 4800 Miles
From China to California
From California to China
Sandwich Is.

C. Guardafui
C. Comorin
Mocha
Sunda Str.
BORNEO
BATAVIA
JAVA
Banda Sea
Arafura Sea
GUINEA
Torres Str.

C. Amber
MADAGASCAR
Mauritius
Bourbon or I. of France
C. St. Mary

I N D I A N   O C E A N
O C E A N
Tropic of Capricorn

O C E A N
E Q U A T O R
P O L Y N E S I A
Marquesas
Society Is.
Breadth of Ocean 9850 Miles

New Caledonia
Norfolk I.

A U S T R A L I A
BRISBANE
PERTH
Adelaide
SYDNEY
MELBOURNE
Bass Str.
AUCKLAND
Tasmania
HOBART TOWN
NEW ZEALAND

PRINCIPAL MOUNTAINS

ASIA
No.	Name	Feet
1.	Gourianshar	26,862
2.	Dhawalaghiri	26,000
3.	Jawalura	25,500
4.	Hamar	21,000
5.	Mauna Roa	18,970
6.	Ophir	13,847

AMERICA
No.	Name	Feet
1.	Aconcagua	23,944
2.	Chimborazo	21,464
3.	Cotopaxi	18,875
4.	Mt. St. Elias	17,900
5.	Sierra Nevada	16,785
6.	St. Helena	15,760
7.	Popocatepetl	17,800
8.	Rocky Mts.	13,567

EUROPE
No.	Name	Feet
1.	Mt. Blanc	15,740
2.	Mt. Rosa	15,172
3.	Mt. Cervin	14,765
4.	Shreckhorn	13,386
5.	St. Bernard	11,018
6.	Simplon	11,000
7.	Mt. Etna	10,963
8.	Col du Cervin	10,502

AFRICA
No.	Name	Feet
1.	Geesh Mts.	15,058
2.	Kilimandjaro	11,900
3.	Atlas Mts.	12,048
4.	Tenerife	12,738
5.	Lamalmon	11,700

ASIA   AMERICA   EUROPE   AFRICA

PRINCIPAL RIVERS

NOTE.—The Southern Ocean embraces the cold oceanic regions south of the Pacific, Atlantic, and Indian Oceans. The portion of it within the Antarctic Circle, is sometimes called the Antarctic Ocean.

LESSON XIV.
Studies on Mercator's Map of the World.

What great land-mass occupies a central position on this Map?—Where does it approach nearest to another great body of land?—Name the six largest divisions of land, including the largest island.—Which of these are partially represented on both the Eastern and Western sides of the Map?—How many Oceans are there?—Name all the Oceans.—Where is Cape Horn?—Cape St. Roque?—Cape Race?—Strait of Belle Isle?—Cape Farewell?—Point Barrow?—Cape Prince of Wales?—Cape Blanco?—Cape of Good Hope?—Cape Guardafui?—Cape Comorin?—The Strait of Sunda?—Tasmania?—The Sandwich Islands?—Kamtchatka?—Nova Zembla?—North Cape?—Lofoden's Islands?—Cape St. Vincent?—Cape Verde?—The Canary Islands?—The West Indies?—Spitzbergen?

What Ocean lies east of America?—What lies west?—What division of the Earth lies east of Europe?—What Ocean west?—What Ocean lies south of Asia?—East of Asia?—North of Asia?—How are North and South America united?—What Ocean east of Africa?—What west?—What Sea north?—What is the greatest breadth of the Pacific Ocean?—What of the Atlantic?—Where is the Atlantic narrowest?—How wide is it there?—*The term Antarctic Ocean is sometimes applied to the ocean supposed to exist south of the Antarctic Circle.*

Through what parts of the world does the Arctic Circle run?—the Tropic of Cancer?—the Equator?—the Tropic of Capricorn?—the Antartic Circle?—What great cities lie near the 40th parallel of north latitude?—the 50th?—Through what parts does the 30th parallel of *south* latitude pass?—Through what parts of the Earth does the meridian of Washington run?—the meridian of Greenwich?—the meridian of Peking?—that of San Francisco?

Bound North America—South America—Asia—Europe—Africa—Australia—Spitzbergen—Arctic Ocean.

Where is the Caspian Sea?—the Red Sea?—the Great Salt Lake?—Hudson's Bay?—the Great Lakes of North America?—the Mediterranean Sea?—the Black Sea?—the Baltic?—the North Sea?—the Caribbean Sea?—the Gulf of Mexico?—the Gulf of California?

Where is the Mississippi River?—the Amazon?—the Nile?—the Rhine?—the Danube?—the Volga?—the Ganges?—the Amoor?—the Darling?

Where are the Alleghany Mountains?—the Rocky Mountains?—the Andes?—the Himalaya?—the Alps?

Find Washington, New York, San Francisco, Quebec, Liverpool, London, Paris, Madrid, Hamburg, Berlin, Bogota, Lima, Mexico, Florence, Constantinople, St. Petersburg, Yakutsk, Petropaulowski, Honolulu, Melbourne, Havana, Reikiavik, Peking, Yedo.

Which contains the most land, the Northern or Southern Hemisphere? Ans. *The Northern contains three times as much as the Southern*—Which contains most, the Eastern or Western Hemisphere? Ans. *The Eastern contains twice as much.*—Which Zone has the greatest proportion of land? Ans. *The North Temperate Zone: it has thirteen times as much land as the South Temperate Zone.*—How much of the earth's circumference in the equatorial region is water? Ans. *Four-fifths.*—How are the great Peninsulas of the Earth generally projected? Ans. *Toward the south, e. g.,* Spain, Arabia, Indo-China, Corea, Kamtchatka, Africa, South America, Alaska, California, Malacca, and Greenland, and several other peninsulas.—How far have geographers explored the Northern Hemisphere toward the North Pole? Ans. *Not further than the 82d parallel of latitude.*—On what parallel of latitude are both the Old and New World broadest? Ans. *On the 50th parallel of north latitude.*—Considering Australia a Continent, what proportion of the Earth's

known land do you suppose consists of islands? Ans. *About one-twenty-fifth.*—How do the mountain ranges of the New World run?—How do those of the Old World run, mostly?—In and near what Zone do you find the highest mountains? Ans. *The Torrid.*—Compare the coasts of North and South America with those of Greenland, Europe, and Africa, and see if they would *fit into each other,* if brought together.

NOTE.—Does not the eastern angle of South America look as if it had been torn out of the Gulf of Guinea, and the western projection of Africa out of the Gulf of Mexico? You see how the projections of one coast correspond with the recesses in the opposite coast; even the mountains and plains of the one correspond with those of the other.

The meridian of Tenerife, one of the Canary Islands, divides the Earth into two parts; in one of them the land greatly predominates, in the other the water predominates. On which side of that meridian is the great mass of land?

LESSON XV.
About Climate.

1. Climate is the combined effect of light, heat, electricity, and moisture, and is manifested in what we call " the weather."

It was held for a long while that the climate of a place depended only upon its latitude, but places in the same latitude may and often do have very different climates.

2. Mountain Tops.—It is a well known fact that the weather on the top of a mountain is generally cooler than the weather at its foot. Indeed there are some parts of the Rocky Mountains, even in our own country, uninhabitably cold. They are always covered with snow, both in winter and summer.

There are, in South America, where the Andes are crossed by the Equator, peaks, such as Antisana, which, although it is a burning volcano, pushes its top up to the height of 19,137 feet above the level of the sea—so high as to reach frozen regions. On the top of that mountain, even in the Torrid Zone, with the sun directly overhead, the cold is bitter.

The line of elevation above which the cold is, at all seasons of the year, sufficient to congeal the moisture of the air and form snow, is called *the Snow-Line.*

3. The Snow-Line at the Equator.—The limit of the snow-line at the Equator is 16,000 feet above the sea; so there are mountains in the Torrid Zone on the tops of which the weather is as cold as it is at the North Pole. The ice never thaws, and no green thing can grow there. In descending these snow-capped mountains in the Tropics, we experience, in a ride of a few hours, all the changes of climate that would be felt in travelling from Spitzbergen to Cuba in a single day.

4. Causes which influence climate.—From facts like these, it appears that climate depends upon height above the sea, as well as upon distance from the Equator. But climates are influenced by other circumstances also.

The Island of Great Britain and the Territory of Labrador lie between the same parallels of latitude, and yet, notwithstanding the Highlands of Scotland and the mountains of Wales, are higher than the hills of Labrador, the winter climate of England is so mild that the pastures are green all the year, and London has to depend upon the ponds of New England, which are six degrees nearer to the Equator, to fill its ice-houses; while the winter climate of Labrador, on this side of the Atlantic, is so cold as to render the country uninhabitable.

Scotland, in the northern part of Great Britain, is not only more mountainous than Labrador, but it is also farther from the Equator; yet its winter climate is very much milder than that of Labrador.

This difference of climate depends both upon the situation of those countries with regard to the sea, and upon the prevailing direction of the winds.

In the British Islands, the winds come from the ocean. They are loaded with moisture from the sea, and warmth from the Gulf Stream. In Labrador they come from the land, and are dry and cold. It is owing to the direction of the winds with regard to the sea that the climates of Oregon and British Columbia are mild like those of Western Europe, not cold like those of countries in the same latitude on the eastern coast of America.

5. A Rule for Climates.—In countries within the temperate zones, where the prevailing winds come from the sea, the climates are not so cold in winter nor so warm in summer as they are where the winds come from the land.

The prevailing direction of the wind throughout the temperate zones is from the westward, but from the eastward in the torrid zone.

Notwithstanding the mild winter climate of Great Britain, the summer there is too cool for Indian corn to grow.

Thus, in order to judge properly as to the climate of a country, the geographer has to take into consideration not only its distance from the Equator and its height above the level of the sea, but its distance from the sea and the prevailing direction of the winds also, and ascertain whether they come from the land or from the water.

The table-land of Mexico is six or eight thousand feet above the sea; and notwithstanding it is in the torrid zone, it is not so high as to have a cold, or so low as to have a hot, climate. There are only two seasons there. We have four. With the Mexicans, the year is divided into the rainy and the dry seasons; they have no wintry weather. The cactus, with its variety of forms, is the characteristic vegetation of the Mexican table-land.

6. Isothermal Lines.—To give a better idea of climates than a knowledge merely of latitude would convey, Humboldt introduced the plan of drawing in the maps lines through all places having the same average temperature. These lines are irregular curves, and they are called Iso-therms (same temperature).

They do not tell the climate, but they convey a better idea of what the climate of this or that place probably is, than a mere statement of its latitude or elevation above the sea would do.

The striking *bend* of these isotherms is due to the presence of moisture in the air, the agency of the winds, and of the great currents of the Ocean.

The isotherm of New York, immediately on leaving the Atlantic coast, is bent down by a cold Arctic current running near the coast; but as soon as this line enters the Gulf Stream it inclines northward toward Europe, and comes out on the other side of the Atlantic, 1500 miles farther north than New York.

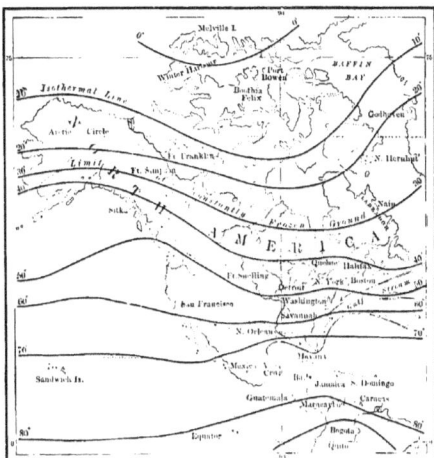

ISOTHERMAL LINES.

7. Industries and Climates.—Though the geographical distribution of all *agricultural* labor is almost wholly an affair of climate, there are other industries, such as mining and manufacturing, that are, to a certain extent, independent of climate.

The laboring man in some countries abandons the cultivation of the soil, especially in those parts of the world where the sea with its bounties, or the factory with its attractions, or the forest with its game, becomes more tempting than the soil.

8. Rule for Labor.—From all this we derive this geographical rule: As you recede from the warm climates of the South and approach the cold regions of the North, human labor becomes less and less agricultural, and the occupations of man more and more diversified.

It is in obedience to this law that the Northern States are more devoted to manufactures and commerce and seafaring than the Southern States are.

Questions.—What is climate ?—Does the climate of places depend entirely upon latitude and elevation above the sea-level ?—What other two conditions influence climate ?—As you ascend a high mountain does it grow warmer or colder ?—What is the snow-line ?—How high above the level of the sea is the snow-line at the Equator ?—How high is Antisana ?—Compare the climate of England with that of Labrador.—They are in the same latitude ; show the contrasts.—Point out these countries on the map. (Mer-

cator's.) Can you describe the difference in their climates ?—How do you account for that difference ?—How does the climate of Oregon and British Columbia compare with the climates of countries between the same parallels of latitude that front on the Atlantic, as Maine and Labrador ?—What is the general direction of the prevailing winds in the temperate zone ?—What in the torrid ?—What are the conditions upon which climates mainly depend ? —What is the height of the table-lands of Mexico ?—Describe the climate there.—Does Indian corn grow in England ?—What are isothermal lines ?— Who suggested them ?—Can you trace the isotherm of 40° ?—How many degrees of latitude does it cross in running from the west to the east coast of the United States ?—Is there any geographical reason why the people of the New England States should, more readily than the people of the Southern States, resort to the sea, the railroad, and the factory for a living ?—Point out on the map those countries whose sea-coast lies within the temperate zone. —What geographical rule does this lesson teach with regard to human labor?

II. DESCRIPTION OF COUNTRIES.

ANIMAL LIFE OF NORTH AMERICA.

LESSON XVI.

1. *The Discovery of America.*— America is our country, and it is both interesting and instructive to learn its geography before it was inhabited by the white man.

Though America is one of the grand divisions of the world, as late as four centuries ago it was unknown to the people of Europe. It was then inhabited by red men and wild beasts. It was discovered, you remember, by Christopher Columbus in 1492, and was named America after Americus Vespucius, one of his companions, a Florentine.

The first land discovered was what is now called Watling Island, which is one of the small islands of the West Indies. This event led to the discovery of other islands, and finally of the American continent itself.

2. The West Indies.—Columbus at first thought that the islands he had discovered were the EAST Indies, and when he found out they were not, they were called the *West* Indies. In soil, climate, and productions there is a striking similarity between the two groups.

3. Indians.—Both North and South America, with their adjacent islands, were inhabited by Indians, of whom only the Peruvians and the Aztecs of Mexico were civilized.

ALPACA LLAMA.

4. Domestic Animals.—There were no horses and no milch kine on the Continent at the time of its discovery. Neither was there any draught animal, beast of burden, or domestic animal of any kind known to the natives, except the hairless dog of Cuba and the llama of Peru.

5. Sargasso Sea.—Among the strange things that Columbus and his men came across on their outward voyage was, as you have learned in a previous book, the Sargasso Sea. That same sea of weeds is still there. It embraces an area of thousands of square miles.

6. The Trade-Wind.—Another thing that amazed and alarmed his crew was that, after leaving Spain and getting out to sea, they found the wind did not change its direction. Day after day it continued from the northeast. They feared that they should never be able to return to their own country against such a wind. This was the northeast trade-wind, well known to every sailor of the present day.

7. Strange Things in the New World.—But as wonderful and as marvellous as these things were, the red men of America, with their pipes, bows, and tomahawks; the inter-tropical vegetation of the islands, with its delicious fruits and beautiful flowers; and the perpetual summer of the Tropics, with its soft climates, were still more so; and the stories told by the simple natives of mighty nations and golden treasures in a land still farther to the westward, filled the minds and inflamed the imagination of these daring mariners with the most dazzling pictures and extravagant allurements—such

as a fountain in Florida, whose waters imparted perpetual youth to all who bathed in them; and a King in the fastnesses of South America, who every morning was anointed with oil and covered with gold-dust. This was the famous but fabled El Dorado (*gilt with gold*), who lived somewhere on the banks of the Orinoco, in a city called Manoa. Its houses were roofed with gold, and its streets paved with precious stones.

8. Spanish and Portuguese Settlements.—Spain followed up her splendid discoveries in the New World by immediate possession, and the prompt establishment of colonies, of which Cuba and the Spanish Republics of America are the living memorials.

She was closely followed by Portugal, who afterwards claimed Brazil by right of supposed priority of discovery.

Bartholomew Dias, one of her navigators, doubled the Cape of Good Hope six years before Columbus saw America. Another, Vasco de Gama, made the first voyage to India by that route.

9. The Line of no Variation.—Between these two powers the New World was partitioned according to a supposed physical, but an ill-defined and ever-changing line—*the line of no variation*. In Europe the needle points to the west of north. Columbus, as he crossed the Atlantic, was the first to discover this *variation ;* and the Pope, who was then at the height of his power, used this line of no variation as a line of geographical division between the possessions of the two wrangling powers. But this line is ever changing its position, and failed entirely to subserve the purposes of a geographical boundary.

10. Early Colonies.—The first colony was established on the James river at Jamestown, Virginia, in 1607. That was soon followed up by the establishment of other colonies. The Roman Catholics settled Maryland, the Quakers Pennsylvania, the Puritans New England.

The Dutch had settlements in New York, the Swedes in New Jersey, and the Danes in Delaware; but in 1662 the inhabitants of these colonies became subject to the rule of England.

San Augustine (*San au-gus-teen'*), in Florida, is the oldest settlement in the whole United States. Florida belonged to Spain until 1819, and the Spaniards had a settlement there—at San Augustine—nearly fifty years before Capt. John Smith established at Jamestown England's first American colony.

Questions.—*1.* On what Continent do you live?—When and by whom was America discovered?—By whom was it inhabited?—Whence does it derive its name?—What was the first land discovered? *2.* How did the West Indies get their name? *3.* Were any of the aborigines of America civilized?—Who? *4.* What was their only domestic animal and beast of burden? *5.* What and where is the Sargasso Sea?—Point it out on the map. *6.* What natural phenomenon alarmed the crew of Columbus?—Why? *7.* What things in the New World appeared most astonishing to the discoverers? —What fabulous stories did they hear and believe?—Where was the Fountain of Youth?—Who was El Dorado? *8.* What Nation was the first to plant colonies in the New World?—Why did Portugal claim a portion of it?— Who was her great navigator, and what discoveries did he make?—By whom and how were the rival claims settled? *9.* What is the line of " no variation?"—Is it stationary? *10.* Which is the oldest city in the United States?— How long after Spain established settlements in America before England began to plant colonies in the New World?—When and where was her first colony established?—What were the other early settlements?

4

NORTH AMERICA

LESSON XVII.

Studies on the Map of North America.

Boundaries.

Within what Meridians and Parallels is North America included?—How is North America bounded?—What body of water connects the Atlantic with the Arctic Ocean?—What Strait connects the Arctic and the Pacific Oceans? (see p. 10).—What Sea and Gulf wash this Continent on the southeast?—What great body of water in the northeastern part of the Continent?—What land lies east of Baffin Bay?—Is Greenland anywhere united to North America?—*Ans.* Nowhere, except by the ice in Smith's Sound.

Where is Smith's Sound?—In what direction does Greenland extend?—Point out Cape Farewell.—Where and what is Iceland?—Point out Cape North in Iceland.

What Cape is first seen by a vessel crossing from England to America?

Indentations.

How and where does the Atlantic Ocean make indentations into the shores of North America? *Ans.* In the Gulf of Mexico and the Gulf of St. Lawrence.—Where does the Pacific indent the Continent? *Ans.* In the Gulf of California.—Does not the Gulf of Mexico look as if the Ocean had scooped out the land? The great Equatorial current, coming from the Eastern side of the Atlantic, enters the Gulf of Mexico between Yucatan and Cuba, and sweeping in a circuit of the Gulf, issues at the Florida Channel, in the Gulf Stream.

Configuration.

Through what Strait would you enter Baffin Bay from the Atlantic?—How would you enter the Arctic Ocean from the Pacific? (p. 10).—Point out Yucatan (*yu-ca-tan'*).—Where is the Bay of Honduras?—Point out the West Indies.—Point out Cape Sable—Cape Hatteras—Cape Cod—Cape Race—The Strait of Belle-isle—Hudson Strait—Lancaster Sound—Barrow Strait—Melville Sound—Cape Bathurst—Point Barrow—Fox Channel—Behring Strait—Melville Island—Queen Charlotte Sound—Cape Flattery—Cape Mendocino (*men-do-see'-no*)—Point Conception—Cape San Lucas.

Water Divisions.

Find Lake Nicaragua—Great Salt Lake—Lake Superior—Lake Michigan—Lake Winnepeg—Lake Athabasca—Great Slave Lake.

Find Ballenas Bay—Chesapeake Bay—Ascension Bay—Gulf of Tehuantepec (*teh-wan'te-pek*)—Bay of Panama.—Where is the Gulf of Darien?—Where is Campeche Bay?—Where is James Bay?

Mountains.

What three mountain-ranges are there in North America?—Do they run in the general direction of the sea-coasts near them?—How do the Rocky Mountains run?—The Sierra Nevada?—The Alleghany Mountains?—Which is the longest range?—Which the shortest?—Which the loftiest?—Which contains Mount Shasta? The Rocky Mountain range crosses the Isthmus of Panama and is prolonged in the Andes of South America to the extremity of that continent at Cape Horn.—How long is that? (See Map, p. 20.)

Where are the Sierra Madre Mountains?—The Cascade Mts.?

Point out Mount St. Elias. This is the highest peak in North America.—Where is Mount Hecla?—Long's Peak?—Pike's Peak?—Spanish Peaks?

How are North America and South America united?—How wide, by the Map and Scale, is the Isthmus of Darien in its narrowest part?

Islands.

Name some of the principal islands near the Arctic coast of the Continent—near the Atlantic coast—near the Gulf of Mexico—near the Pacific coast.—Point out Cuba—Hayti—San Domingo—Porto Rico—Andros Island—Newfoundland—Vancouver Island—Queen Charlotte Island—Long Island—Prince of Wales Land—Breton Island—Anticosti.

Rivers.

Where is the Mississippi River?—Trace its course—Where, the Missouri?—Trace it.—The St. Lawrence—the Ohio—the Ottawa—the Cumberland—the Brazos—the Rio Pecos—Lewis River—Clark's River—Red River—the Arkansas—the Tennessee—the Platte—the Saskatchawan—the Mackenzie—the Yukon—the Columbia—the Colorado—the Rio Grande?—How long is the Mississippi?—The Missouri?

Where does the Rio Grande del Norte rise?—In which direction does the Mississippi flow?—the Ohio—the Saskatchawan—the Yukon—the St. Lawrence—the Mackenzie?

Peninsulas.

Point out the Peninsulas of Yucatan, Florida, Nova Scotia, Alaska, and California.—Victoria Land, King William's Land, Boothia, Melville Peninsula.

Political Divisions.

What three great countries lie wholly north of the Tropic of Cancer? What are the boundaries of the United States?—What, of Alaska?—Of Labrador?—Of the Dominion of Canada?—Of Mexico?

Bound Mexico.—What country lies between the Caribbean (*kar-ib-be'an*) Sea and the Pacific Ocean?

To what power do Cuba and Porto Rico belong? *Ans.* To Spain.—To what power does Jamaica belong? *Ans.* To England.—Is Hayti independent? *Ans.* Yes.

Miscellaneous.

In which zone is North America mostly situated? Where would you find tropical productions on the Continent? Where are the finest fisheries? *Ans.* On the northeastern and north-western coasts.

Where is grain most extensively cultivated? *Ans.* In the Valley of the Mississippi.

Where are the winters coldest, on the coast or in the interior of the same latitude? *Ans.* In the Interior.

Which are the two largest political divisions of North America?—Which is next in size?—Next?—Which is the smallest?

Are all the West India Islands in the Torrid Zone?—How much of Mexico is in the Temperate Zone?

Where do the Esquimaux live?—Where is Upernavik?—Hopedale?—St. John's?—Havana?—Aspinwall?—City of Mexico?—Chihuahua (*che-wah'-wah*)?—San Francisco?—Virginia City?—St. Paul?—Detroit?—Buffalo?—Albany?—Washington?—Memphis?

What mountains and rivers would you cross in going directly from Washington city to San Francisco?—From the city of Mexico to New York?—Where is the Magnetic Pole?

What are the most northern lands of North America?—the most eastern?—the most western?—the most southern?

What parallels of latitude would you cross in going from Panama to Point Barrow?—What meridians of longitude would you cross in going from Sitka to St. John's, Newfoundland?

LESSON XVIII.

Our own Country—Continued.

1. Early English Settlements.—After establishing the colony at Jamestown, in Virginia, England proceeded to establish and acquire others, until her colonies numbered thirteen. These were Virginia, Maryland, Georgia, North Carolina, South Carolina, Delaware, Pennsylvania, New Jersey, New York, Massachusetts, Rhode Island, New Hampshire, and Connecticut. These are often called the "*Original Thirteen.*"

These colonies were governed very much in the same way as the Dominion of Canada now is, but had no voice in their own government.

2. The Revolution.—After many years of unwarranted restrictions and painful exactions, they became dissatisfied with the treatment they received from the mother country, and complained to the King and his ministers. The colonists obtained no redress of their grievances. Then, in the persons of their representatives, they met in what is now called "Independence Hall," at Philadelphia, on the 4th of July, 1776, and declared themselves free, sovereign, and independent. But it became necessary to fight, to make the declaration good.

Finally, after seven years' war, we were acknowledged by Great Britain herself to be thirteen independent powers. Afterward the "Thirteen" agreed to unite under the Constitution, and create the government of *The United States of America.*

3. Political Changes, etc.—At the end of the Revolutionary War, Kentucky, and all that territory east of the Mississippi, which now constitutes the States of Ohio, Indiana, Illinois, Michigan, and Wisconsin, belonged to Virginia. She ceded it to the government of the United States in trust for the benefit of all the States alike.

In 1803 Louisiana, which then included what is now Arkansas, Missouri, and other States west of the Mississippi, was purchased from the French.

In 1819 Florida, which belonged to the Spaniards, was purchased from Spain.

In 1845 Texas—having previously revolted from Mexico and established her independence—was annexed to the United States, and became one of them.

In 1848 California was bought from Mexico for $20,000,000, and Arizona, in 1854, for $10,000,000. Finally, in 1867, Russian America, marked on the maps

as Alaska, was purchased from Russia for $7,200,000, and thus the borders of the "Original Thirteen" have been enlarged to their present gigantic proportions.

Questions.—1. Name the original thirteen. To whom did they belong as colonies?—How were they governed? *2.* How did they become separated from the British Crown?—When and where did they proclaim their independence? *3.* What was the extent of the Territory that originally formed the State of Virginia?—What did she do with it?—How have the United States since enlarged their borders most, by purchase or by conquest?

----•----

LESSON XIX.

Geographical Position and Features of America.

1. Geographical Position.—Before we go farther into the political geography of our own country, let us take a general survey of Mercator's map (pp. 21 and 22), and study the geographical position which the American Continents occupy with regard to the rest of the world. This position is important in its commercial, political, and social aspects.

2. Extent.—The Continents of America, as you will observe on the map, stretch from the frozen regions of the North to the inhospitable climes of Cape Horn (*Cabo de Ornos*), so called from the number of burning volcanoes which the Spaniards, who first doubled it, saw.

They called it, therefore, the Cape of the Furnaces or Ovens (Ornos).

These Continents extend from the parallel of 75° North to that of 57° South latitude, which is more than nine thousand statute miles in length; and they include within their borders large portions of four out of the five great zones into which geographers have divided the earth. Area, excluding islands, 15,000,000 sq. ms.

Asia and Africa lie each partly in three, and Europe in two of the zones; and though America is not the largest continent in area, it includes a greater range of latitude, embraces a greater diversity of climate, and yields as great a variety of vegetable productions as all the others put together.

3. Advantages of Position.—America lies between the two great Oceans; it has the Atlantic on the East and the Pacific on the West. Its harbors on the Pacific are midway between the western shores of Europe and the eastern shores of Asia.

With its double sea front, numerous harbors, narrow isthmus, and central position, combined with its great length from North to South, and the quick transit across it, and its great diversity of climate, America is destined to be the most maritime and commercial of all the continents.

Commercially, it is in the position of a half-way house between the maritime nations of Western Europe and

Eastern Asia. Highways for commerce have been constructed across it, and more are in preparation.

Australia is a British Possession, and already the Isthmus of Panama is made the thoroughfare for mails and passengers between that land of gold and its mother country.

The maritime nations, and especially the United States, are making serious efforts to open a ship-canal from the Atlantic to the Pacific. Should this be accomplished, through the Isthmus of Tehuantepec, or Panama, the canal would doubtless become the great commercial thoroughfare for the world's caravan of trade.

4. Rivers and Lakes.—America possesses larger rivers than any of the other continents. It is more abundantly watered than any of them, and it has a smaller extent of desert waste and barren land than either Africa or Asia, and is capable of sustaining a larger population than either.

More than half of all the fresh water in the world is contained in the Great North American Lakes, and there are no rivers anywhere of such volume as the Amazon and the Mississippi.

What drains and gutters are to the streets, and ditches to the farmer's low grounds, rivers are to the country at large—they collect the drainage and carry it off.

The proper study of the rivers and the mountains and the coast-lines of a country, simply as they are delineated on the maps, is most instructive.

5. Running Water.—The face of the earth has been made what it is by the influence of mountains, the agencies of rivers, and the action of water. Travel through the country where you will, and you will see in the rounded pebbles, in the layers of rocks and soil, or in the distribution of sand, evidence of this action.

Wherefore, in studying a river in its geographical aspects, you are not to consider so much the width of its channel, or the depth and capacity of its current, as you are to study the offices which the water, as it rolls along, performs, and the extent of country to which it gives drainage.

6. Further Offices of Rivers.—More clearly to understand the offices of rivers in the economy of nature, let us follow in imagination the waters which feed them from the time they come from the sea as vapor until they return to it again through the river. It has formed clouds to decorate the sky and screen the earth from the heat as well as cold. It has been condensed into rain, and refreshed the land with showers. It has filled the water-veins in the earth which feed the springs and wells, and in doing that it has collected food for the inhabitants of the sea, for it has dissolved and worn away the rocks, and torn off from them the materials of which sea-shells and coral are made; and while it was doing all this, it turned mills, drove machinery, floated ships, and carried the rich produce of our land to market.

7. The Watershed of a river is formed by the sides of the hills and valleys which slope toward it, and from which it receives the drainage. The valley, or hydrographic basin, of a river, is the whole extent of country that is drained by it. Thus we speak of the valley of the Mississippi, or the hydrographic basin of the Amazon, and mean all that part of the continent from which the waters run into those rivers.

The valleys of these rivers form the two largest hydrographic basins in the world, and as the roofs of the largest houses require the largest gutters to carry off the water, so the largest hydrographic basins require the largest rivers to drain them—the rain-fall being the same.

Thus, simply by looking at the map and tracing out by the eye the various river basins, you arrive at the conclusion that the Amazon and the Mississippi are the largest rivers in the world; and so they are.

So, also, you might judge of the extent of an unexplored river basin by ascertaining the volume of the river through which its water is discharged.

8. Mountains.—America has also the longest range of mountains in the world. The Andes take their rise in Patagonia, and skirting the Pacific coast of the country, extend all the way to the isthmus which connects North and South America, stooping down to hills of two or three hundred feet in height. They rise up again, as you go north, until they reach the grand proportions of the Sierra Madre—as the range is called in Mexico—and then of the Rocky Mountains in the United States, the name by which they are known in all their northern extent.

THE OROGRAPHIC VIEW OF THE UNITED STATES, on the next page, reveals to you at a glance our great mountain-systems, our valleys and hills, the large divides and watersheds, which assist in giving direction, volume, and velocity to our noble rivers ; it also shows our immense prairies and slopes—in a word, it presents us with a miniature pictorial model of the face of our country. Examine it long and carefully.

The different tints of light and shade represent the elevations and depressions of the surface of the country ; the dark shades and tints show the low lands and deep valleys; the lighter tints show the higher lands and the mountains.—In the darker shades rivers are marked by a white line; and in the lighter shades by a black line.

Questions.—1. Why is our geographical position important ? *2.* What are the northern and southern boundaries of the American Continent ?—What is the distance between them ?—How did Cape Horn get its name ?—How many Zones does the Continent of America cross ?—Which of the four Continents has the greatest diversity of climate ? *3.* Which coast of America is nearly midway between the western shores of Europe and the eastern shores of Asia ?—What are the geographical circumstances which are destined to make this Continent the most commercial and maritime of the four ? *4.*—What gives America its great diversity of climate and variety of productions ?—What two Oceans bathe its eastern and western shores ?—Which of the Continents is most abundantly watered ?—Which has the least Desert ?—Why is this Continent capable of sustaining a larger population than any of the others ? How does it compare with them as to the size of its Lakes and Rivers ? *4.* What proportion of the fresh water in the world do the Great Lakes contain ?—What, in a geographical point of view, is the use of Rivers ? *5.* Name some of the offices of water.—Did you ever see any evidences of its action ? *6.* Tell what the water in a River has been doing since it left the Sea as vapor. *7.* What is the Watershed of a River?—A Hydrographic Basin ?—Trace out with your finger on the map the Hydrographic Basin of the Amazon—the Valley of the Mississippi—the Ohio—the Tennessee—the Columbia—the Potomac—the Orinoco—the La Plata.—How, by looking at the map, can you judge as to the relative size of Rivers ?—Why does a large Watershed require a larger River than a small one to carry off the water from it ?—Where are the two largest Rivers in the world ? *8.* Which Continent has the longest mountain range ?—Trace it out on the map.—To which coast is it nearest ?—What names does it assume ?

On which side of the OROGRAPHIC VIEW OF THE UNITED STATES do you distinguish the loftiest mountains ?—On which side, consequently, do you find the longest and most majestic rivers ?—How is the Mississippi Basin formed ? *Ans.* By the slopes of the Alleghany Mts. on one side, and of the Rocky

ATLANTIC OCEAN

GULF STREAM

GULF OF MEXICO

OROGRAPHIC VIEW
OF THE
UNITED STATES

JAMES BAY

L. ONTARIO

L. ERIE

L. HURON

L. MICHIGAN

LAKE SUPERIOR

GULF OF CALIFORNIA

PACIFIC OCEAN

Mountains on the other side.—Do you discern in the Orographic View a gentle elevation south of Lake Erie? From the crest of this, some streams run northward into the Lake, and some southward into the Ohio River.

Why are the streams west of the Rocky Mountains smaller than those on the eastern side of the range?—How are the great Lakes fed?—Judging by the eye, which has the most rain, the eastern or western side of the Alleghanies? Ans. The western.—Why is this? Ans. Because the eastern side of the mountains gets little rain from the sea; the prevailing winds are from the west.

LESSON XX.

A General View of the Geography of the United States.

1. Extent.—The United States extend from the Atlantic to the Pacific Ocean, and the distance from sea to sea, in a direct line across the country, is 2,100 miles in the narrowest and 2,600 in the broadest part.

In latitude the United States extend from the Great Lakes on the north to the Gulf of Mexico on the south. Further to the west they stretch from the confines of Mexico to the British possessions on the north.

From their extreme northern to their extreme southern limits they embrace, including the newly-acquired territory of Alaska, forty degrees of latitude; but the choicest parts of the United States lie between the parallels of twenty-six degrees and forty-nine degrees.

2. Climate and Productions.—This breadth of latitude, as you may have inferred from the preceding lesson, is the cause of a great diversity of climates in this country, with their varieties of production.

This portion of the North Temperate Zone corresponds in climate and productions to that which is occupied in Europe by the most enlightened and prosperous nations; but Europe is not blessed with such a variety of climate and productions as we have here. Our great staples of cotton, rice, sugar, and tobacco cannot be cultivated to a profitable extent in Europe.

All the great agricultural staples of commerce, including tea, coffee, and indigo, are to be found in the United States.

The climate of Florida and the lands bordering the Gulf of Mexico is semi-tropical. To the north of this belt we have, either on the plains or in the valleys or on the mountain slopes, all climates that are to be found in Europe between the frozen regions of the North and the sunny plains of Greece and Italy.

3. Our Sea-Fronts and Harbors.—On the east, south, and west we look out on the sea, and American waters are more lavish with their bounties than those of other countries. They abound in excellent harbors, in which our neighbors are deficient.

Nature has placed in the way of Mexico and the Central American States obstacles which bar them, in a measure, from the industries of the sea, and tend to place obstructions in their way as great maritime and commercial powers.

Nor are our neighbors on the north more favored in this respect than our neighbors on the south, for in the dominion of Canada the harbors are closed annually with ice, and navigation suspended for many months; and when the harbors are free, their offings are often beset with icebergs, and made dangerous to the navigator by reason of dense fogs; whereas the harbors and coasts of the United States are ever free and open.

4. Inland Navigation.—Inland, and midway between the Atlantic and Pacific, the Mississippi River flows from north to south, receiving richly-laden vessels from its navigable tributaries as they pour in from the east and west, and giving to the inhabitants of its valley a free outlet to the sea for their produce and merchandise.

The great Lakes, from Superior to Ontario, are inland seas, which have the commerce of an ocean and greatly facilitate trade and intercourse between the people of the neighboring States.

5. Nearness to Market.—With all these advantages of geographical feature and position, we are nearer to the markets of Europe than are the people of either China or Japan; and consequently can undersell them there.

With a canal through the Isthmus we shall also be nearer to the markets of China, Japan, and the East than the merchants of Europe are.

Thus you perceive that this country occupies geographically the most favored position.

Questions.—1. What is the greatest distance in a straight line across the United States between the Atlantic and Pacific Oceans?—What the least?—Across which tier of States does it run, the Northern or the Southern?—How many degrees of latitude do the United States, with the newly-acquired territory of Alaska, embrace? **2.** To what is the great diversity of climate in the United States attributable?—Why are the United States able to boast of a greater variety of production than Europe?—Which of our great agricultural staples cannot be profitably cultivated there?—What great staples of commerce are cultivated in the United States?—In what parts of the United States is the climate semi-tropical?—In the other parts what are the climates? **3.** Why, commercially speaking, is the geographical position of the United States more advantageous than that of any other continental power?—Compare the United States with the countries immediately to the North and South with regard to harbors.—Why are Mexico and the Central American States never likely to become great maritime and commercial powers? **4.** Can you name any other circumstances besides those which you have already mentioned, which tend to make the geographical position of the United States so favorable? **5.** Why should American merchants be able to undersell European merchants in the markets of China and Japan?

LESSON XXI.

Studies on the Map of the United States.

Configuration and Boundaries.—Point out on the map, the broadest and the narrowest part of the United States.—Between what parallels of latitude do the United States lie?—Between what political divisions do they lie?—How are the United States separated from the British provinces?—Name the four great Lakes that lie between the United States and the Canadas.—Name the Rivers or Straits that connect these Lakes one with another.—Is there any natural boundary between the United States and Mexico?—What?

The boundary line, you observe, runs through the middle of these Lakes and Rivers, thus showing that they belong equally to England and to us; and that the navigation of them is alike free to both nations.

Mountains.—Point out the principal mountain chains between which the Mississippi valley lies.—On which side of this valley are the Rocky Mountains?—On which the Alleghanies?

Rivers.—Where does the Mississippi River rise?—Which way does it flow, and where does it empty?—Measuring in a straight line, according to

THE
UNITED STATES
OF
AMERICA.

POSSESSIONS

ONTARIO

QUEBEC

JAMES BAY

L. of the Woods

Rainy L.

I. Royale

LAKE SUPERIOR

LAKE MICHIGAN

LAKE HURON

NEW BRUNSWICK

MAINE

ST. LAWRENCE RIVER

QUEBEC

MONTREAL

OTTAWA

NEW HAMPSHIRE

VERMONT

MASSACHUSETTS

NEW YORK

ALBANY

BOSTON

PORTSMOUTH

CONCORD

AUGUSTA

ST. PAUL

MADISON

MILWAUKEE

CHICAGO

DETROIT

CLEVELAND

BUFFALO

PENNSYLVANIA

HARRISBURG

PHILADELPHIA

IOWA

DES MOINES

ILLINOIS

SPRINGFIELD

INDIANAPOLIS

COLUMBUS

OHIO

ST. LOUIS

CINCINNATI

WHEELING

PITTSBURG

VIRGINIA

RICHMOND

MARYLAND

DELAWARE

WASHINGTON

TOPEKA

JEFFERSON CITY

KENTUCKY

LOUISVILLE

LEXINGTON

FRANKFORT

RALEIGH

NORTH CAROLINA

RALEIGH BAY

C. HATTERAS

ALBEMARLE SOUND

Springfield

TENNESSEE

NASHVILLE

KNOXVILLE

MEMPHIS

LITTLE ROCK

ARKANSAS

SOUTH CAROLINA

COLUMBIA

CHARLESTON

ATLANTA

AUGUSTA

GEORGIA

MILLEDGEVILLE

SAVANNAH

MISSISSIPPI

JACKSON

ALABAMA

MONTGOMERY

LOUISIANA

BATON ROUGE

NEW ORLEANS

MOBILE

PENSACOLA

TALLAHASSEE

FLORIDA

JACKSONVILLE

ST. AUGUSTINE

Tampa Bay

C. Romano

C. Sable

C. Florida

GULF STREAM

BAHAMA ISLANDS

Little Abaco

Great Abaco

SAN SALVADOR

ATLANTIC OCEAN

GULF OF MEXICO

RED RIVER

Shreveport

Galveston Bay

GALVESTON

Corpus Christi Bay

MATAGORDA

the scale on the map, what is the length of the Mississippi from north to south?—Through how many degrees of latitude does it flow?—Which is its largest tributary from the west?—What from the east?

You observe that the Mississippi River is less crooked than the Missouri; the latter flows east through Montana into Dakota and then southeast to its juncture with the Mississippi at St. Louis. Measuring in like manner in a straight line from the source to the bend in Dakota, and then to its mouth, how long is the Missouri? From its junction up, which drains the largest valley, it or the Mississippi?—Which drains the largest valley, the Ohio or the Mississippi—above their junction?

You observe in California an inland range of mountains, the SIERRA NEVADA (snowy mountains), with a COAST RANGE in front, flanked in Oregon and Washington by the CASCADE RANGE, and that the valley between the Nevada range and the Rocky mountains is very broad. In this valley lies the *Great Inland Basin* of the Continent. It has no sea-drainage, and in it the Great Salt Lake and other lakes of like quality are found.

Remember this as a general rule : *All lakes that have no outlet are salt.*

What rivers rise west of the Rocky Mountains?—Which way do they flow?—Which of them is the longest, measuring in a straight line from source to mouth?—Repeat the general rule about salt lakes.—Why do you call a part of the valley between the Sierra Nevada and the Rocky Mountains an inland basin?—What is the meaning of Sierra Nevada?

Bays, Straits, Gulfs, Capes.—Name the bays along the Atlantic shores of the United States; the straits ; the gulfs ; the capes ; the peninsulas.—Where are the Bahama Islands?—What large river drains the great lakes?—What large river flows into the Gulf of Mexico?—Describe the course of each.—What range of mountains runs parallel with the Atlantic coast ? Name the gulfs and bays along the Pacific coast of North America.—Name the straits, capes, and islands.—Name the two largest American rivers that empty into the Pacific Ocean.—What chains of mountains run parallel with the Pacific coast?

Name the principal islands and capes along the Arctic shores of North America.—The bays, sounds, and gulfs.

What large river empties into the Arctic Ocean?—Describe its course.—Which is the largest, British America or the United States, judging by the eye ?—Mexico or the United States?—Which of these three divisions abounds most in lakes and islands, and indentations along the sea-shore ?

States.—Beginning with the most northerly, name, in order, the thirteen Atlantic States and their capitals.—Mention in the same way the States bordering on the Gulf of Mexico.—The States and territories on the Pacific.—Name the eight States on the Great Lakes.—What State is partly bounded by the St. Lawrence river?—What States touch British America?—What territories lie contiguous to Mexico ?

Routes.—What States and rivers would you cross in going directly from New York to Chicago ?—What States, territories, and rivers would you cross in going directly from New York to San Francisco ?—From Washington to Charleston ?—Should you take a steamboat at Pittsburg for St. Louis, on what rivers and near what States would you pass?—In going from Pittsburg to New Orleans by steamboat, what States would you see ?—Should you sail from Portland, in Maine, to Galveston, Texas, what States, bays, and gulfs would you pass?—Sailing from Alaska to San Francisco, what capes, islands, and States would you pass?—How would you go by steamer from New York to San Francisco ? *Ans.* By the way of Aspinwall, where you would cross the isthmus on the Panama Railroad, and take another steamer from Panama to San Francisco.

Through what river-valley would you pass in going by steamboat from New Orleans to St. Paul ?—How would you go to Omaha by steamboat from New Orleans?—How, by the same means, would you go to Mobile ?—To Louisville?—To Fernandina, Florida ?—To El Paso, Mexico ?—To Austin ?

LESSON XXII.

The Political Subdivisions of the United States.

1. The Census.—A census is taken by the general government of the United States once every ten years. By the census of 1860, the population was 31,443,321 ; by that of 1870, it was nearly 39,000,000. This makes the United States more populous than any of the European nations, except Russia and Germany.

The population of Great Britain is 37,000,000; of France, 35,000,000; of Russia, in Europe, 69,000,000; and of the new German Empire, 45,000,000.

2. States, Territories, etc.—The United States consist of thirty-eight States and ten Territories, which are variously divided into groups or sections for the convenience of reference.

3. Divisions.—The ordinary grouping of these thirty-eight States and ten Territories, is this :

1st. The six *New England*, or *Eastern States*,—Maine, New Hampshire, Vermont, Massachusetts, Rhode Island, and Connecticut.

2d. The five *Middle States*,—New York, New Jersey, Pennsylvania, Delaware, and Maryland, with the District of Columbia.

3d. The eleven *Southern States.*—Virginia, North Carolina, South Carolina, Georgia, Florida, Alabama, Mississippi, Louisiana, Texas, Arkansas, and Tennessee, with the Territories, New Mexico and Indian Territory.

4th. The thirteen *Western States,*—West Virginia, Kentucky, Ohio, Michigan, Indiana, Illinois, Missouri, Iowa, Wisconsin, Minnesota, Nebraska, Kansas, Colorado, with the Territories, Wyoming, Dakota, and Montana.

5th. The three *Pacific States,*—California, Nevada, and Oregon, with the Territories of Washington, Idaho, Utah, Arizona, and Alaska.

By a more natural classification, we might group the States as the Atlantic, the Gulf, the Inland, the Lake, and the Pacific States.

We also speak of the Valley States, meaning those that are in the Mississippi Valley ; or the Cotton States, meaning those in which cotton is the principal staple, as Texas, Louisiana, Mississippi, Alabama, Florida, Georgia, and South Carolina.

Questions.—1. What is the Census? What was the population of the United States in 1870 ?—How does it compare with the population of England ?—How with that of France ?—of Russia? *2.* How many States and territories are there ? *3.* How many and what sectional divisions are there? —Name the States and Territories of each section.—What other groupings are sometimes made?—Point out on Map of United States all the above-named States and Territories.—Bound them in order.

LESSON XXIII.

The New England States.

TOTAL POPULATION, 3,487,924.

STATE.	Capitals.	Chief Cities and their Population.
Maine	Augusta	Portland 31,413
New Hampshire	Concord	Manchester..... 23,536
Vermont	Montpelier	Burlington 14,387
Massachusetts	Boston	Boston 250,526
Rhode Island	Providence / Newport	Providence 68,904
Connecticut	Hartford	New Haven 50,840

1. Map Study.—The most lively impressions as to the geography of a country are to be obtained by travelling over it ; but this cannot always be done, and the best ideas that we can obtain as to all except the Political Geography of a country, are to be obtained from the map. Study the maps ; look at them often and attentively, and you will soon get the sections of the country fixed in your mind as indelibly as your playground and gymnasium.

Let us, therefore, begin to study, with the map before us, the geography of the six New England States.

They are situated in the northeast corner of the United States, between the parallels of 41° and 47° 20′ north latitude. They are nearer to Europe than any other part of the United States. They are small States. The six put together are not as large as Missouri, nor one-third the size of Texas ; but, according to area and population, they have more power in Congress than any other part of the United States twice their size. They are bounded—See Map—on the west by the State of New York, on the north and east by the Dominion of Canada, and on the south by the Atlantic Ocean. Rhode Island is the smallest State of the Union.

Rhode Island has two Capitals.

2. Coast Indentations.—One of the first things that strike the eye of the geographer, as he turns it for the first time upon a map of these States, is the very jagged appearance of their coast-line, especially of Maine, and the number of rocky islets which curtain the shores—sure signs that there is no lack of deep water and good harbors.

The New England States are, as you might therefore infer, rich in harbors, bays, capes, and islands, for this is shown by a glance at the map.

Vermont is the only one of the New England States that has no sea-front.

3. Lakes and Rivers.—Another striking feature also most prominent in Maine, is the number of fresh-water lakes that dot its surface, as well as the number of small streams which, in all the States of this section, thread their way from the hills to the sea.

Some of these rivers flow west into Lake Champlain, some north into the St. Lawrence ; but most of them run east and south, and empty into the Atlantic Ocean.

Thus the New England States, in their orography, are like a " hipped-roof house" which sheds the water off in four directions, their largest watershed sloping toward the Atlantic Ocean.

4. " Orography" means the irregularities of the earth's surface up and down, or in the vertical way. You have already seen an Orographic View of the United States. (p. 30.)

5. Mountains.—Now when we remember that, after those of North Carolina, the highest mountains in the " Atlantic States" are in the New England States (Mount Washington, White Mountains, New Hampshire, 6,234 feet ; Mount Mitchell, a peak of the Blue Ridge, North Carolina, 6,770 feet high), and when we consider how close the mountains and highlands of New England are to the sea, we may understand that the streams which flow thence to the sea have a great descent, with rapid currents and falls sufficient to afford abundance of water-power.

6. Facilities for Mills and Factories.—Such is the case : and the New Englanders recognizing the force of those principles which regulate the geographical distribution of labor, have erected mills, factories, and manufacturing establishments along the banks of these streams. They have invoked the aid of this power, and made it subservient to their purposes.

7. Manufacturing Towns.— Fall River, Lowell, Lawrence, Manchester, Springfield, Nashua, with many other places, are celebrated for their water-power and for their manufacturing industry.

The New England States have no coal-mines, and their hills are poor in metallic ores.

8. Climate.—Their winter climate, though they front on the sea, is very cold. The sea there, however, is never frozen, and the harbors very seldom closed, not because the weather is not cold enough, but because the sea is so deep and the tides so rapid. Therefore, even in the severest weather, its waters, on account of their depth, are comparatively warm.

Because the waters are warmer than the air, the coast of New England is not so cold as the interior.

You know that the water from a deep well or a good spring is cool in summer, warm in winter. It only *appears* so because in winter the weather is cold and in summer it is warm, while the temperature of the water in the well or spring is nearly the same in winter as in summer.

It is for this reason that the winds which blow over large bodies of water are cool in summer, warm in winter.

9. Prevailing Winds. — The prevailing winter winds in New England are from the land, not from the sea. They are consequently dry and cold ; the weather, therefore, is often bitterly severe. But when the wind comes from the sea, as it sometimes does, it makes damp, foggy, and disagreeable weather, especially in winter, late autumn, and early spring. Moreover, the winters there, besides being severe, are, by reason of the latitude, long, while the summers are short. Consequently the soil yields scantily, and agriculture is by no means the most profitable branch of industry that the inhabitants of a country so favored with water-power, so blessed with harbors, so convenient to the sea with its bounties, and so rich in timber for ship-building, may pursue.

10. Resources and Industries.—The agricultural labor of New England does not yield food enough for home consumption. But they gather abundant harvests from the sea and its fisheries.

With the severe climate and stingy soil on one hand to make agriculture uninviting, and with their forests of ship-timber, their capacious harbors, their water-power, and sea-fisheries to attract on the other, it is no wonder that the sons of New England should have resorted to these industries, for they find in them ample rewards for the labor and hardihood which they demand.

11. Fisheries. — Marblehead. Newburyport, and Gloucester are the chief towns that are engaged in the fisheries ; they are all three in Massachusetts, and their fishing-grounds for cod and mackerel are on the banks of Newfoundland.

These they salt and dry, and then bring them home, and afterward send them to all parts of the world — especially to the Roman Catholic countries, where the people during Lent eat no other animal food.

Thus the people of the New England States, consulting the geographical position of their section in connection with its natural resources, have made lumbering, shipbuild-

ing, ice-harvesting, seafaring, and manufacturing their most important industries.

Questions.—1. What can you learn from maps?—How large are the six New England States ?—Where are they situated ?—Between what parallels of latitude do they lie?—How are they bounded ?—Which is the smallest State in the Union ?—How does their power in Congress compare according to and in proportion with other parts of the Union ? *2.* What is the most striking feature presented by the map of these States?—Which of these States has the longest sea-coast and the best articulated coast-line ?—What geographical conclusion do you derive from these indentations of shore-line ? *3.* What striking feature in the land do you observe on the map of the New England States ?—To what can you liken the watersheds presented by these States ?—Which way do they slope ?—Where do the rivers from them empty ? *4.* What is the *Orography* of a country ? *5.* What, where, and how high is the loftiest mountain of New England ?—Is it the highest mountain in the Atlantic States ? *6.* What effect have the hills of New England and their distance from the sea upon the industry of these States ? *7.* What and where are the most famous manufacturing towns of New England ?—How do you account for its mild coast climate ? *8.* Does the sea freeze off the coast of New England ? *9.* Why are the winters in New England long and the summers short ? *10.* Can you explain why the industries of New England are rather commercial and manufacturing than mining ; seafaring, than agricultural ? *11.* Name three of the principal fishing towns in New England.—Point them out on the map.—Where are their principal fishing-grounds ?—Where are their fish markets ?—What are the chief industries of the New England States ?

LESSON XIV.

The New England States—Continued.

MAINE.

Maine excels in shipbuilding, the lumber-trade, and the harvesting and export of ice. The timber is cut during the winter, hauled in the frozen snow to the banks of the streams, upon which it is launched and

ICE-HARVESTING.

floated down in the spring. Most of it is drifted down the Penobscot river to Bangor, whence it is exported ; but much of it finds its way to Bath and other seaport towns which are celebrated for their lumber-trade and shipbuilding. The fine ships that they send thence to help carry on the commerce of the world, have spread the fame of the New England shipwrights along the sea-coasts of all countries.

Portland is the principal seaport of Maine. It is in railway connection with Canada. A vast amount of the direct trade between Canada and England is carried on through Portland. Emigrants and travellers from England to Canada often, especially in winter, go through Portland, and frequently through Boston.

New Hampshire.

This State has only a few miles of sea-front, and her people are not given to seafaring so much as those of Maine and Massachusetts, which have extensive lines of sea-coast.

The people are extensively engaged in her quarries of stone, and in the manufacture of cotton and woollen goods. Manchester, Nashua, Dover, with other towns, prosper in this business.

Portsmouth, on the Piscataqua (*pis-kat'a-kwah*) river, is the only seaport town in New Hampshire. It has a splendid man-of-war harbor, upon which, just across the border, at Kittery in Maine, the United States have one of their finest navy-yards. Hanover is the seat of Dartmouth College.

Massachusetts,

Rhode Island, and Connecticut are more extensively engaged than any other States in manufacturing. More than one-third of all the woollen, cotton, and leather goods that are manufactured in the United States are manufactured in these three States.

Boston is the commercial emporium and pride of New England. This famous city is situated on a fine harbor in Massachusetts. Yielding to New York in commercial importance, it boasts of rich merchants and much capital, and vies with Philadelphia and Baltimore in foreign trade.

The largest and most celebrated manufacturing establishments in Massachusetts are at Lowell and Lawrence, famous for their woollen and cotton goods. The United States have their most extensive armory at Springfield, where muskets and other small arms, for the public service, are made. Lynn is celebrated for shoemaking.

New Bedford and Nantucket are largely engaged in the whale and other deep-sea fisheries.

Rhode Island.

Newport, in Rhode Island, is situated on one of the finest harbors in New England. It is a noted watering-place, where the sea-bathing in summer is exceedingly fine. The largest manufacturing establishments in Rhode Island are in Providence, where there are extensive cotton factories.

Connecticut.

Connecticut is especially famous for the manufacture of small wares, such as clocks, sewing-machines, pistols, hooks and eyes for ladies' dresses, etc. There are large establishments in Bridgeport for the manufacture of sewing-machines, and one in Hartford for the manufacture of fire-arms.

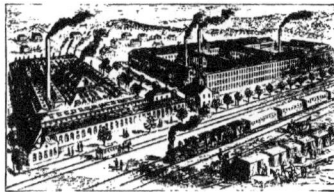

SEWING-MACHINE WORKS.

Yale College, at New Haven, Harvard in Massachusetts, Princeton in New Jersey, and William and Mary in Virginia, are the oldest colleges in the country.

New London and Stonington, on Long Island Sound, are engaged in whale-fisheries.

Vermont.

Vermont is an inland State ; it is therefore cut off from the sea and its marts ; consequently the industrial pursuits of Vermont are more or less different from those of her five sister States, for they can be neither commercial, seafaring, nor fishing.

In Vermont, grazing seems to be the most important branch of industry. Vermont is famed for her horses, and has fine quarries of marble and slate. Both she and New Hampshire are also fine wool-growing countries.

Note.—Common schools are more general in New England than they are in any other section of the United States.

Their best endowed and most renowned institutions of learning are: Harvard University, Cambridge, Mass.; Yale College, New Haven, Conn.;

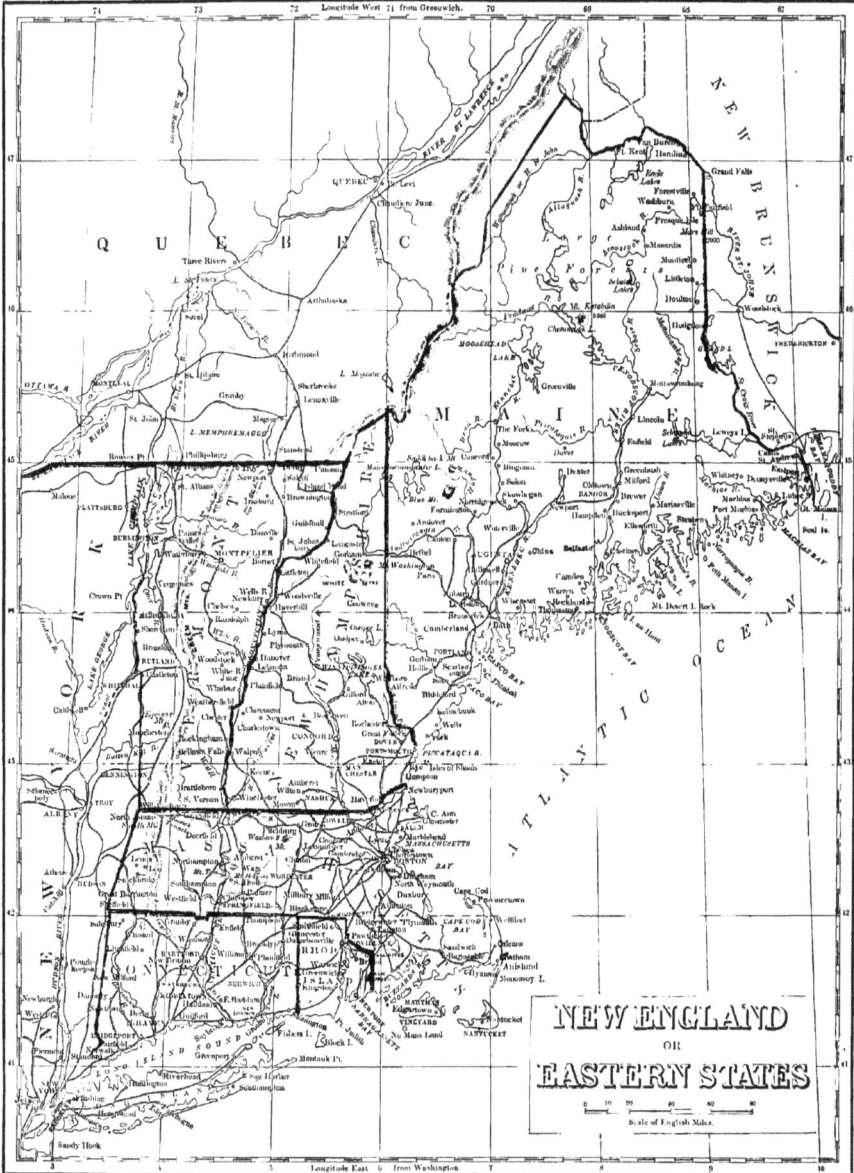

NEW ENGLAND
OR
EASTERN STATES

Scale of English Miles.

Brown University, Providence, R. I.; Middlebury College, Vt.; and Dartmouth College, N. H. The University of Vermont is in Burlington.

Questions.—What is the principal branch of industry in Maine?—At what season of the year is the timber gathered?—What are the chief seaport towns in this State?—What, the shipbuilding places?—Where is Portland?—What business is done through it?

What are the principal branches of industry in New Hampshire?—Why are they not connected with the sea as extensively as Maine and Massachusetts?—How does the coast-line of New Hampshire compare with theirs as to extent?—How many seaport towns has New Hampshire?—What great public establishment is situated on Portsmouth harbor?—Point out some of the principal manufacturing towns in this State.

Which three of the New England States are most extensively engaged in the business of manufacturing?—What and where is the commercial emporium of New England?—Where are the most extensive manufacturing establishments in New England?—Name the principal wares that are manufactured at each.

In what industries are New Bedford, New London, and Nantucket largely engaged?—Where is Newport, and what is it famous for?—For what class of wares is Connecticut especially celebrated?—Name the oldest institutions of learning in the United States.—Where are they?—Why do the industries of Vermont differ from those of the other New England States?—For what class of industries is Vermont most noted?—Name the chief colleges and universities in New England.—Tell the capital and chief city of each State.—Population of chief city in each State.

LESSON XXV.

Studies on the Map of New England.

Boundaries.—How are the New England States bounded?—Bound each one separately.—Which one has the greatest length of sea-coast?

Water Divisions.—Name the principal bays, capes, and islands along the coast.—What Rivers have their rise in the New England States, and their mouth in the British possessions?—What, judging by the eye and according to the valleys, are the largest Rivers in Maine?—Which is the longest River in the New England States?—Where does this River rise?—Which way does it flow?—Where does it empty?—What River rises in the White Mountains?

Mountains.—*The White Mountains are a cluster of high peaks in New Hampshire, of which Mount Washington is the highest. They are an offshoot or spur of the Green Mountains, in Vermont, which are a branch of the Alleghanies, that take their rise in Georgia and skirt the Atlantic sea-board all the way until they end in the Green Mountains of Vermont, and are cut off by the St. Lawrence in Canada.*

NOTE.—The Rivers of New England, for reasons already explained, have a rapid fall, and therefore none of them are navigable any great distance, although the tides rise to a great height along the New England coast.

Trace the irregular line which, in the New England States, divides the watershed of the Atlantic from the watershed of the St. Lawrence.

The Divide.—*You observe that all the Rivers on one side of this line flow for the most part in a southwardly direction, while those on the other side flow northwardly.*

This line, you must remember, is called the "DIVIDE" between the two watersheds.

NOTE.—You discover, also, that it is among the valleys in the upper and middle parts of these watersheds, that Lakes most abound.

Lakes serve an important office as reservoirs or cisterns for receiving the water during floods, or in periods of heavy rains, and distributing it gradually, by evaporation in times of drought.

Look at the map and tell which are the largest Lakes. Which of the New England Lakes has outlets to the sea through the St. Croix? The Penobscot? The Kennebec? The Androscoggin? The St. Johns? The Merrimac? The Richelieu (*ree-she-lu*)? The St. Francis? Where are the two last named rivers? To which watershed do they belong?

What lake borders on Vermont and New York?—What is the distance from the head of Lake George to the head of Richelieu River?

Capitals.—Name the Capitals of each of the New England States.—Tell upon what river, and upon which bank of the river—right or left—they are situated.

(*Remember, the right or left bank of every river is the right or left hand side of that river as one descends it.*)

What is the course and distance of each one of these Capitals from Boston?—How far, and in what direction, is each Capital from the centre of the State?

Chief Towns.—What is the chief town in each of these States?—Describe its situation.—Where is Cambridge?—Bath?—Bangor?—Lawrence?—Lowell?—Newburyport?—Lynn?—Gloucester?—Marble Head?—Springfield?—New Bedford?—Nantucket?—Tell what each one is noted for.

Routes for Travellers.—How would you go from Boston to Lowell?—From Boston to Hartford?—To Lynn?—To New Bedford?—To New London?—To Manchester?—To Montreal?

How would you go from Providence to New Haven?—From New York to Brattleboro?—From Hartford to New York by steamboat?—From Concord to Boston?—From New York to Rutland by railroad?—From Manchester to Providence?—From New York to Boston by water?

Questions.—Can you explain any use of lakes in physical economy?—What is the right bank of a river?—Are the rivers of New England navigable for great distances?—Why not?—Of what mountains are the White Mountains a spur?—Of what are the Green Mountains a branch?—Where do the Alleghanies begin?—What is the highest peak of the White Mountains?

LESSON XXVI.

The Middle States.

TOTAL POPULATION, 9,848,255.

STATE.	Capitals.	Chief Cities and their Population.
New York	Albany	{ New York..... 942,292 { Brooklyn 396,099
New Jersey	Trenton	Newark..... 105,059
Pennsylvania	Harrisburg	Philadelphia..... 674,022
Delaware	Dover	Wilmington 30,841
Maryland	Annapolis	Baltimore 267,354
District of Columbia		Washington 109,199

1. Geographical Features. — Always keeping the map before you, let us now proceed to gather from it some idea as to the principal geographical features of those five States, with the District of Columbia. (p. 45.)

You see that they are traversed by the Alleghany

Mountains, which run from southwest to northeast, nearly parallel with the coast, and that these mountains are divided into ridges, which lie parallel with each other ; the distance between the top of the eastern and the western range varying from 50 or 60 to 100 miles, or more.

These mountains divide these States into two grand watersheds, one of which slopes toward the southeast, and carries the drainage off into the Atlantic Ocean ; and the other to the northwest, with drainage both for the great Lakes, by numerous small streams, and for the Gulf of Mexico through the Ohio and Mississippi rivers.

You observe, therefore, that the Middle States lie partly in the Mississippi valley, partly in the basin of the great Lakes, and partly along the Atlantic slopes.

You notice, also, by the rivers, that there are here and there gaps in the mountains, through which the waters break and pass from one side to the other

Thus the head-waters, both of the Delaware and the Susquehanna, rise beyond what appears to be the crest of the Alleghanies, and passing through a gap in these mountains, find their way to the Atlantic Ocean.

These water-gaps, as they are called, are found in all mountainous countries ; and sometimes the scenery about them is very beautiful, wild, and grand.

2. Position.—The Middle States lie between the parallels of 38° and 45° ; they embrace seven degrees of latitude, and extend several degrees farther to the south than the New England States do ; consequently their climates are milder, their agricultural productions more varied, and many of their chief industrial pursuits different.

3. Size.—New York is the largest and Delaware the smallest of the Middle States.

New York alone is three-fourths the size of all the New England States put together ; while all the Middle States together are about half the size of Texas.

4. Coast-Line.—The coast-line of the New England States—especially of Maine—is, as you remember, of rock, while that of the Middle States is chiefly of loam.

5. Alluvial Country.—In one section the sea is encroaching upon the land, and has worn the shore away to the solid rocks ; in the other the land is encroaching upon the sea, and is gaining upon it continually.

The seaboard of the Middle States, and of all the country to the south, is formed in part of what the sea has cast up, and in part of what the rivers have brought down from the mountains. It is therefore an *alluvial* country. Every rain muddies the rivers, and these muddy waters flow into the sea; there the mud settles and gradually forms land. A large portion of the best and richest countries of the world are *alluvial* ; that is, the soil has been brought down

little by little, by waters from the hills. The meadows and low grounds along the margins of the brooks and streams are alluvial.

6. Alluvial Country in the United States.—The extent of this alluvial country in the United States increases as you go south ; following the coast-line until you get to the mouth of the Mississippi, it extends far up on both sides of that river, and embraces large portions of Illinois, Indiana, and Ohio. The coral, the sea-shells, and other marine fossils found there, show that there, also, the sea once rolled its waves. (Map, p. 32.)

The inland limits of tide-waters are marked by falls or rapids, as those of the Schuylkill at Philadelphia, of the Patuxent near Baltimore, of the Potomac at Georgetown, of the Rappahannock at Fredericksburg, of the James at Richmond, of the Roanoke at Weldon, and so on along the whole Atlantic seaboard.

7. Falls.—These falls are at the head of navigation and of tide-water, and the belt of country between them and the sea is called in each State the *tide-water* or low country.

This belt, though it increases in breadth as you go south, is not so broad as it would be had the rise and fall of the tides been as great in the South as they are in the North.

8. Tides.—In the Bay of Fundy, which borders the coast of Maine, the rise and fall of the tide is sixty feet, whereas, as you go south and reach the shores of the Carolinas, it is only as many inches ; and when you get to the shores of Florida, in the Gulf of Mexico, it is not so much as a foot.

9. Climate.—The Gulf Stream, with its tepid waters, sweeps close to the shores of the Middle States.

The New England States have high and cold mountains from which the west winds blow. The Middle States, on the contrary, New York especially, have the lakes to the westward of them, and they temper in winter the keen west winds as they sweep by. Such is the influence of lakes and large sheets of open water in mitigating the severities of winter climates.

In sheltered spots upon the borders of Lake Ontario it is so mild that even the peach will mature in the open air. The cold west winds, after crossing the mountains and being chilled, make the climates of New England entirely too severe for this delicious fruit.

10. Pursuits.—The difference of climate in these two sections is also both seen and felt in its effects upon the industrial pursuits of the people.

Consequently there is less of seafaring and manufacturing in the Middle States, and far more agriculture and mining industry. The former States are poor in mines, but the latter are rich in both coal and iron.

11. Products. — Wheat, rye, oats, and Indian corn, buckwheat, orchard-fruits, berries, and garden vegetables, all do well in the Middle States.

Still, so great is the town and city population of these States in comparison with their rural population, that as you pass in review State after State from Maine toward the South, you find none of them producing corn and wheat enough for their own consumption until you come to Maryland. She is the first State that regularly produces enough and to spare. She also grows more tobacco than she requires for her own use, and sends large quantities of it abroad.

Questions. —Name the Middle States. *1.* By what range of mountains are they traversed? —What are their principal geographical features as shown on the map?—Describe the great watersheds into which they are divided by these mountains. *2.* Where do the Middle States lie?—Between what parallels?—What rivers have water-gaps through the Alleghanies? *3.* Which is the largest and which the smallest of these States? *4.* Describe their coast-line. *5.* Where in the Middle States is the alluvial country? *6.* Describe it.—Describe the tide-water country and the navigation of the rivers.—How are its limits marked? *7.* Contrast the tides and the shore-line of the New England and the Middle States. *8.* How high does the tide rise in the Bay of Fundy?—How high in the Gulf of Mexico? *9.* Which of these two sections has the mildest climate?—What is the cause of this difference? *10.* What are the industries of the Middle States? *11.* What are the agricultural productions of the Middle States?—What, the mineral productions? —As you travel south along the seaboard from Maine, which is the first State you come to that produces breadstuffs enough for home consumption?

SCENE IN BROADWAY, NEW YORK.

LESSON XXVII.
Middle Atlantic States.

NEW YORK.

The climate of Western New York is tempered by the lakes, which soften the west winds as they sweep over them. That part of the State is, therefore, a fine wheat and corn country. It is also a good grazing country, and the hardier orchard-fruits do well there. New York is also a wool-growing State.

There are salt-springs at Syracuse which are owned by the State : the water is sold to the salt-makers, who produce annually about one-sixth of all the salt that is consumed in the United States.

Apart from her salt-works and stone-quarries, her beds of gypsum, and some iron ores near the Pennsylvania line, New York is poor in minerals.

The city of New York is the emporium of trade for the whole country, and the largest city in the New World.

Albany is one of the most important inland towns in the Middle States, and is, besides, the capital of the most populous State in the Union. It is situated at the head of navigation, on the right bank of the Hudson and at the mouth of the Erie canal, where vessels from the Lakes freighted with western produce meet those that ply up and down the river from New York.

Lake Erie, as is shown by the falls and rapids of Niagara, is more than three hundred feet above Lake Ontario. The smaller lakes, which give such a charm to the scenery in this part of the State, are situated on the terrace with Lake Erie ; consequently the rivers which from these lakes carry water into Lake Ontario, have, like the Niagara, to leap a precipice in order to escape from this terrace. Their rapid falls afford fine water-power for mills and manufacturing purposes. Rochester and Oswego have availed themselves of it, and are extensively engaged in milling and manufacturing. Grist is sent them even from Canada and the West.

The falls of the Hudson are near Albany, and Troy avails herself of the power afforded by them, and applies it to various manufacturing purposes. At Watervliet, between Albany and Troy, the United States have an arsenal. The Military Academy at West Point is on the right bank of this beautiful river, and a few miles below Newburg. There is also an extensive navy-yard at Brooklyn.

The country bordering upon the Hudson is in a high state of cultivation and improvement. The fine houses and beautiful grounds lend enchantment to the scenery, and many travellers take passage in steamers that ply on the river merely for the pleasure of enjoying the beautiful views that meet the eye at every turn. The scenery on the right bank, between West Point and New York, is rendered bold by columns of basaltic rock called the "Palisades," that rise up perpendicularly to the height of four or five hundred feet.

Notwithstanding the number, splendor, size, and fleetness of the steamboats that ply on the Hudson between New York and Albany, a railway has been built directly on the bank all the way between the two cities. The trains run very swiftly, and carry crowds of passengers.

NEW JERSEY.

New Jersey has more sea-front than New York. This State lies almost entirely in the tide-water country. It is not so far north as New York; it has, therefore, a milder climate. New Jersey is famous for its fruit-orchards, and for peaches especially. It is richer in mines of iron and zinc than New York.

There are some flourishing manufacturing towns in this gallant little State. Paterson, at the falls of the Passaic, is celebrated for its railway cars and locomotives—Newark for the extent and variety of its manufactures; and among them, those of india-rubber in particular. Its population has been doubled within the last seven years.

Princeton College is one of the most renowned and ancient seminaries of learning in the United States.

DELAWARE.

Delaware has no mountains and is poor in minerals, but, like New Jersey, it lies mostly in tide-water regions, is rich in soil, and favored with a mild climate. It is a fine grazing country, and the laboring-classes find profitable employment there in supplying the markets of Philadelphia, New York, Baltimore, Brooklyn, and the other large cities and towns in that part of the country with fresh meat, butter, fruit, and vegetables. The most celebrated powder-mills in the United States are on the Brandywine, near Wilmington.

PENNSYLVANIA.

Pennsylvania abounds in coal and iron; the richest wells of petroleum oil are found near Lake Erie, in the upper valley of the Alleghany river, in this State.

The anthracite coal of Pennsylvania is the chief article of fuel that is used for domestic purposes in the Middle States. It is largely consumed, also, for smelting and other manufacturing and mechanical purposes. It constitutes one of the most important branches of trade in the United States. The flourishing cities of Reading and Scranton owe their prosperity chiefly to it.

A COAL-CRACKER.

The quantity of coal exported from this country is very small. All that is mined is required in our domestic economy. Petroleum has suddenly sprung up into great commercial importance; the first oil wells were discovered about fifteen years ago; it is, after cotton and provisions, the chief article of export from this country. It is used mostly for lights, and Germany is our best customer. Cotton is our most valuable article of export. The following is an official statement of the exports from this country for the year ending July, 1871: Cotton, $227,885,000; Breadstuffs, $79,382,000; Oils and Petroleum, $37,313,000; Provisions, $36,444,000; Tobacco, $22,206,000; Naval Stores, $2,704,000; all others, $12,917,000; Total, $412,547,000.

The German farmers of Pennsylvania are noted for their good husbandry and fine barns, which are generally more elegant buildings than their dwelling-houses.

Philadelphia is celebrated for the cleanliness of its streets, its fine market-places, its medical schools, its academies, and its charitable institutions.

In proportion to its population, Pittsburg is more extensively engaged in manufacturing than any other city in the State. Iron and glass are among its chief articles, and it has a large trade, also, in bituminous coal, with which the neighboring hills are bountifully stored.

MARYLAND.

The peninsula between Delaware Bay and the Chesapeake, belongs, a part to Delaware, a part to Maryland, and a part to Virginia. Its elevation above the sea-level is not much : for it is without mountains, and its climate is softened by the mild temperature of the ocean and warmth from the Gulf Stream in winter, and by the cooling sea-breezes in summer.

Moreover, the Chesapeake Bay, with its open water and other benign influences, reaches up into the heart of Maryland, and imparts to the country along its shores, in this State, a softer climate than any other part of the Middle States can boast of.

Tobacco is prominent among the agricultural staples of Maryland, which embrace corn, wheat, and the other cereals.

Its mountains abound in coal and iron, and mining constitutes a large and important branch of the industry of this State.

Baltimore, on the Patapsco river, is the sixth city in the Union.

A SCENE IN BALTIMORE

Cumberland, at the western terminus of the Chesapeake and Ohio canal, is the centre of the mining region.

Frederick, on the border, and Hagerstown, where there is an excellent college, are in the middle of that ex-

ceedingly fertile belt of country, of which the celebrated Valley of Virginia forms a part. Its fertility is owing to the same vein of limestone that makes the Falls of Niagara. The Genesee country, the finest agricultural part of New York ; the Harrisburg country, the best in Pennsylvania ; and the Shenandoah valley the garden spot of Virginia, are all upon this vein of limestone. Weirs' Cave, with its splendid stalactites, near Staunton, and the Natural Bridge in Rockbridge county, near Lexington, with its fine archway, both in Virginia, are formed of this limestone rock.

This vein forks in Virginia, one part passing through Lexington and the Green River country of Kentucky, the other via Abingdon, Virginia, into Tennessee, and so on to the Muscle Shoals of the Tennessee, where the two forks come together and continue their course toward the southwest. It crops out from under the Walnut Hills of Vicksburg; that is the last that is seen of it. The country, all the way from the banks of the Mississippi to those of the Niagara River, is one of unsurpassed fertility.

The blue-grass country of Kentucky and Tennessee, owes its celebrity to this fertilizing vein.

Trace this limestone vein on the map with the eye, for it will enlighten you as to the distribution of labor, and help you properly to understand the geography of your country.

No sheet of water in the world surpasses the Chesapeake Bay for the variety, excellence, and abundance of the fish and game with which its shores and waters abound. Its shad and herring fisheries are very valuable. Seines, a mile long, are hauled for them.

The canvass-back duck, the terrapin, and the oyster of the Chesapeake Bay, are unsurpassed in flavor and excellence. They often grace the tables of royalty in Europe, as delicacies. This sheet of water, on account of the value of its fisheries, has been compared for wealth to the gold mines of California. The oyster-sellers throughout the West are supplied chiefly from its bounties.

DISTRICT OF COLUMBIA.

This District, embracing originally an area of 10 miles square, was ceded, a part by Virginia and a part by Maryland, to the general Government.

In 1846, Congress ceded back to Virginia the portion on the right bank of the Potomac that was originally ceded by her, so that now the District contains only about 60 square miles.

Washington City was planned by General Washington. It is beautifully laid out, on the left bank of the Potomac, and contains many magnificent public buildings, the principal of which is the capitol. It was called District of Columbia after Columbus. It is no longer governed directly by Congress, but, like the territories, has a legislature and one delegate to Congress.

The people of the District have a Governor appointed by the President and Senate.

THE NATIONAL CAPITOL

Georgetown, availing itself of the water-power derived from the Chesapeake and Ohio Canal, is engaged extensively in the manufacture of flour; it has also a considerable trade. This city is connected with Alexandria by the canal which crosses the Potomac at Georgetown, over a magnificent aqueduct.

Washington is without commerce or manufactures; it derives its importance entirely from the presence of the Government.

Questions.—Which State embraces the most degrees of latitude?—What effect have the Great Lakes upon the climate of New York?—What are its principal agricultural staples?—Where are the salt-works?—To what does Albany owe its importance?—How high is the level of Lake Erie above that of Lake Ontario?—How do the streams get from one level to the other?—What use is made of these falls?—What two cities have availed themselves extensively of this water-power?—For what purpose?—Where are the Falls of the Hudson?—What city makes the most extensive use of this water-power?—What important Government establishments are at Watervliet and Brooklyn?—Where is West Point?—What makes it so noted?—Where are the Palisades?—What are they?

Which has the mildest climate and the longest coast-line, New York or New Jersey?—How do you account for this difference of climate?—For what industries are Paterson and Newark noted?—Where is Princeton, and for what is it celebrated?

What are the most valuable minerals in which Pennsylvania most abounds?—What is the chief article of fuel in the Middle States?—Whence is it obtained?—Name the chief article of export—What is the most valuable?—What is the chief market for petroleum?—What can you say of Pennsylvania as an agricultural State?—For what is Philadelphia especially noted?—For what is Pittsburg?—To what do Reading and Scranton chiefly owe their prosperity?—Where are they situated?

What can you say of Delaware?—Name the staple productions of that State, New Jersey, and Pennsylvania.—What famous works are on the Brandywine near Wilmington?—By what States is the Delaware peninsula occupied?—Describe it.—To what do you attribute its mildness of climate?

What are the staple productions of Maryland?—Where are her coal mines?—Upon what vein of rock is Hagerstown?—Look at the map, and name the parts of the Southern States that are situated upon it.—What do you know about the fisheries of the Chesapeake Bay and its tributaries?

What is the Territory of Columbia?—How is its name derived?—Are its inhabitants entitled to representation?—Who makes their laws?—Describe Washington City—Who planned it?—To what does it owe its importance?—Where is Georgetown?—In what business is it chiefly engaged?—How is it connected with Alexandria?

LESSON XXVIII.

Studies on the Map of the Middle States.

Boundaries.—How are the Middle States bounded?—Bound each one, and the Territory of Columbia, separately.

Capitals and Towns.—Give the population of each State, with that of its chief town.—Tell the situation of each capital.—Tell the bearings and distance of each Capital from the chief town of the State.—Which of them extends from tide-water to the Lakes?—Which has the greatest length of sea-coast?

Coast Marks and Indentations.—Name the principal Bays along the coast of the Middle States.

Mention the principal Islands.—The chief Capes.

Mountains.—What mountain system crosses this section of the United States?—In what direction does it extend?—What part of New York does it cross?—Where does it cross New Jersey?—Point out the Adirondacks.—What is the most eastern range of the Alleghany Mountains called?—Name the chains in Pennsylvania, beginning at the eastern part of the State.—Name the mountains in New York.

Falls.—Where are the Niagara Falls?—Trenton Falls are near Utica, N. Y.—Where is Harper's Ferry?—Where is the Delaware Water-Gap? *Ans.* Where the Delaware River forces itself through the eastern ridge of the Alleghanies, in the northwestern part of New Jersey.—What Falls are there at Rochester? *Ans.* Those of the Genesee River, as it makes its way into Lake Ontario.

Comparative Geography.—Look at the map, and describe the difference between the sea-coast of the Middle States and of New England.—Which has the greatest breadth of tide-water country, and the greatest length of navigable rivers, the New England or the Middle States?—On account of what natural cause is it, that the alluvial country in the Middle States is so much broader than it is in New England?—What inference do you draw from this, as to the extent of inland navigation in the two sections?—Name, in each section, some of the principal seaport towns, and describe their situation.

Towns and Cities.—How far is New York from Sandy Hook?—How far from the Gulf Stream? *Ans.* About 210 miles.—Where, in New York, is Syracuse?—Rochester?—Buffalo?—Binghamton?—Ithaca?—Saratoga?—Poughkeepsie?—Brooklyn?

Point out, in Pennsylvania, Pittsburg.—Where is Harrisburg?—Reading?—Scranton?—Mauch Chunk?—Gettysburg?

Where, in New Jersey, is Trenton?—Elizabeth?—Paterson?—Newark?—Princeton?

Where is Dover?—Wilmington?—Where is Annapolis?—Baltimore?—Ellicott's Mills?—Port Tobacco?—Frederick?—Havre de Grace?—Ungerstown?—Georgetown?

Rivers and Lakes.—Describe the course of the Hudson.—Trace the Mohawk.—The Delaware—The Susquehanna.—The Potomac.—Trace the head-waters of the Ohio.—What streams form the Ohio?—Find Lake George.—Lake Champlain.—Where does the St. Lawrence empty its waters?—What lakes are drained by the Oswego River?—What rivers and lakes border on the Middle States?—What State most abounds in lakes?—Upon what watershed are these lakes chiefly situated?—Name the principal rivers of the Southern, the Western, and the Northern watershed.—What tributary of the Ohio takes its rise farthest to the North?—What farthest to the East?—Which is the longest, the Chesapeake or the Delaware Bay?—

QUEBEC

LAKE HURON

GEORGIAN BAY

LAKE ONTARIO

ONTARIO

LAKE ERIE

OHIO

PENNSYLVANIA

VIRGINIA

WEST VIRGINIA

KENTUCKY

NORTH CAROLINA

TENN.

ATLANTIC OCEAN

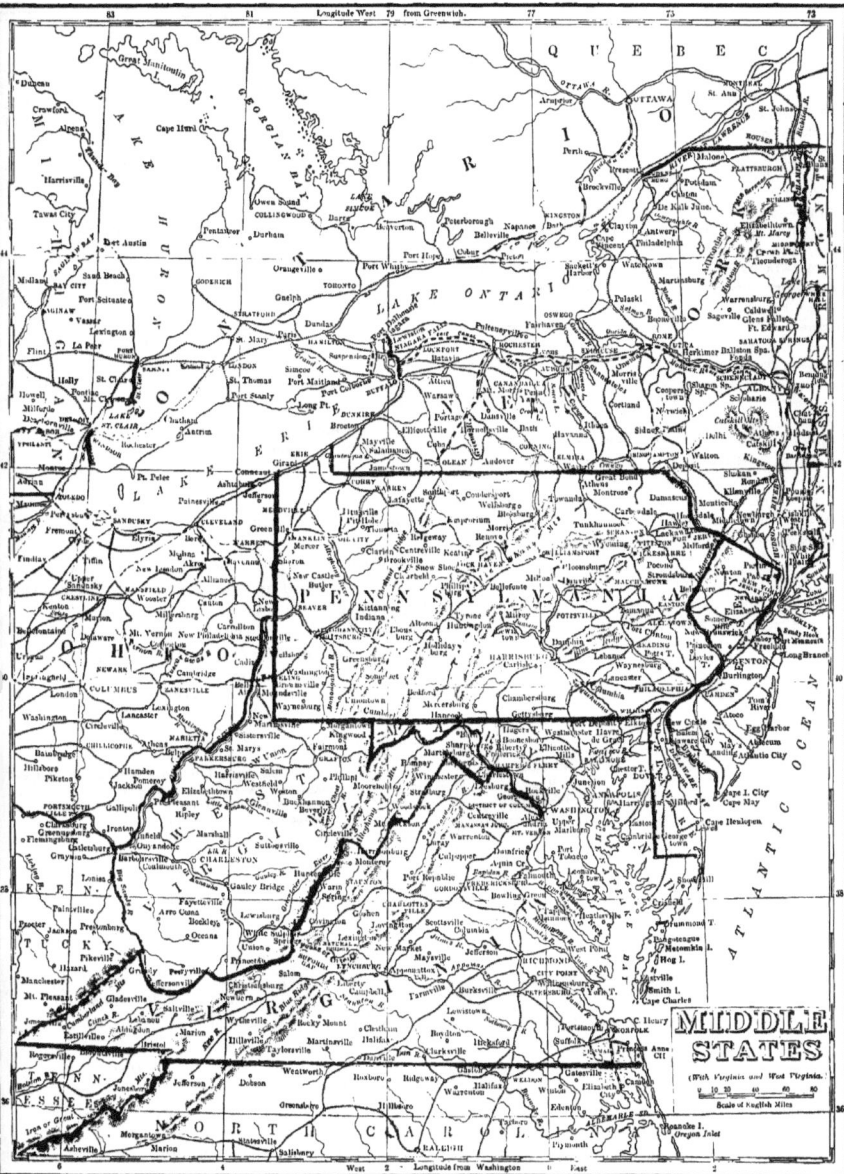

MIDDLE STATES

(With Virginia and West Virginia.)

Scale of English Miles

West 2 — Longitude from Washington 0 East

What Rivers empty into them?—What Capes are at the mouth of each?—What, judging by the eye and the scale of miles, is the greatest length and breadth of each Bay?

Routes and Travels.—How would you go by steamboat from New York to Albany?—On what waters would you sail in going from New York to Philadelphia?—To Baltimore?—How, in the same way, from Buffalo to Oswego?—To Ogdensburg?—How would you go by steamboat from New York to Boston?—From New York to Atlantic City?—To Cape May?—How would you go by railroad from New York to Albany?—To Buffalo?—To Dunkirk?—From Buffalo to Pittsburg?—To Harrisburg?—To Washington?—How would you go by rail from Philadelphia to Pittsburg?—From Wilmington to Trenton?—How would you go from New York, by steamer, to Green Port, Long Island? *Ans.* You would go by way of Long Island Sound.—How do steamers enter the Sound from New York? *Ans.* By passing through East River, an arm or estuary of Long Island Sound, improperly called a river.—Where is Long Branch?—Cape May?—Atlantic City?

Miscellaneous.—To which of the four cities, Boston, New York, Philadelphia, and Baltimore, measuring in a direct line, is Chicago, at the head of lake navigation, nearest? *Ans.* To Baltimore.—To which of these four great Atlantic Cities is Cleveland, Ohio, nearest? *Ans.* To Baltimore.—To which of them is Buffalo nearest? *Ans.* To Baltimore.

On which bank of the Delaware is Philadelphia?—On which bank of the Hudson is New York?—Jersey City?—How is Baltimore situated?—Wilmington?—How is Washington located?—Georgetown?

LESSON XXIX.

The Eleven Southern States and Two Territories.

TOTAL POPULATION, 9,591,260.

States.	Capitals.	Chief Cities and their Population.	
Virginia	Richmond	Richmond	51,037
North Carolina	Raleigh	Wilmington	13,446
South Carolina	Columbia	Charleston	48,956
Georgia	Atlanta	Savannah	28,235
Florida	Tallahassee	Pensacola	3,347
Alabama	Montgomery	Mobile	32,034
Mississippi	Jackson	Natchez	9,057
Louisiana	New Orleans	New Orleans	191,418
Texas	Austin	Galveston	13,818
Arkansas	Little Rock	Little Rock	12,380
Tennessee	Nashville	{ Memphis	40,226
		{ Nashville	25,865
Indian Territory	Tahlequah	Tahlequah	1,000
New Mexico	Santa Fé	Santa Fé	9,699

1. Geographical Position.—The Southern States lie between the parallels of 25° and 40°. These are the most favored latitudes on the earth, both as to climate and production. Between these parallels are found such countries in Europe, as Greece with its Archipelago; Southern Italy and Spain; the Land of Goshen, in Egypt; and in Asia, the Promised Land—the Vale of Cashmere—the Valleys of the Hoang Ho, and the Yang-tse-kiang, where lie the great City of Pekin, and the choicest parts of China and Japan.

Excepting the two countries of Italy and Greece, the Southern States are better watered than any of them.

2. Past and Present Condition.—Domestic servitude was one of the established institutions of the South previous to the recent war. In round numbers there were 4,000,000 of negro slaves, who were emancipated by proclamation from the President of the United States.

In consequence of this and the ravages of war, the industry of the South has been greatly deranged, and the people have not yet had time fairly to adjust themselves to their new situation.

Let us, therefore, content ourselves by looking not so much at the present industries and political condition, as at the natural geography and resources of the Southern States.

3. Peculiar Features.—The Southern States embrace nearly twice the breadth of latitude contained in the Middle States and the New England States together, and they contain four times their area and four times their extent of sea-coast.

The population of the Southern States, however, amounts to only four-fifths of that of the two other sections.

Excepting Virginia, the shores of the Southern States are curtained with a chain of long, narrow, and sandy islands, with navigable inlets and passages here and there obstructed by sandbars which prevent the entrance of ships that draw more than sixteen or eighteen feet of water. The largest ships draw 28 feet.

As we go from the sea inland, anywhere between Virginia and Texas, we cross a belt of swamps, covered with cypress, magnolias, yellow jessamine, and jungle. Then we come to a sandy soil in the piny belt. In some parts these two belts are 300 miles broad; nowhere less than 100. After them come the oaks and the deciduous trees. The cypress belt is noted for its pendent mosses, parasites, and flowers—Magnolia grandiflora, and the yellow jessamine, the loveliest of them all; the pine belt, for its ship-timber and naval stores. In the Gulf States, the cotton produced by this sandy soil, which extends through Mississippi as far up as Tennessee, is called uplands.

4. Watersheds.—The Alleghany Mountains divide the Southern States east of the Mississippi into three watersheds, sloping severally to the east, south, and west, and sending their streams and rivers off into the Atlantic Ocean, into the Gulf of Mexico, and into the Mississippi River.

New Mexico and Western Texas are evidently dry countries, for, as the map shows, they have few watercourses. Consequently, in these regions you would expect to find severe droughts and much barren land.

"Llano Estacado," the *Staked Plain*, in Texas, is an

immense barren waste across which the early Mexican travellers and traders marked their way by sticking up stakes along the trail.

5. Minerals.—The mountainous regions of Virginia, North Carolina, Georgia, and Tennessee are rich in minerals. Iron, coal, lead, copper, salt springs, mineral springs of rare virtue, quarries of marble and gypsum, and veins of gold ; in Louisiana, salt-beds ; in Arkansas, salines and quarries for the best of whetstones ; in Alabama, artesian wells with power to turn machinery :—these are some of the mineral riches and sources of wealth in the Southern States which depend not upon climate.

6. Climate and Occupations.—You already understand enough about climates to infer simply from the map that there is a great difference between the climates of Virginia and the New England States on the one hand, and the climates of Virginia and the Gulf States on the other ; and that, consequently, there should be a corresponding difference in the industrial pursuits.

In Virginia the chief occupations consist in the cultivation of wheat, corn, rye, oats, and tobacco, fruit and vegetables ; in the cutting of firewood and ship-timber for the northern cities ; and in wool-growing, grape-growing, cattle-raising, mining, and fishing.

In the more Southern States, industry, in addition to the raising of flocks and herds, is directed to the cultivation of rice, cotton, and sugar, with breadstuffs and fruits for home consumption, and to the turpentine and lumber business.

7. Influence of Inventions.—Human inventions and improvements are important geographical agents, for they often change or alter the industrial pursuits throughout extensive regions of country. So far as they do this, they bear upon questions, especially of political geography, and they should not escape the attention of those who study this most important and instructive department of human knowledge.

Not only the face of our country, but the chief industrial pursuits of the people have been greatly changed or affected by the invention of the cotton-gin, by the application of steam to machinery and locomotion, and by the various mechanical improvements of the age.

Before Whitney's invention of the cotton-gin, the cultivation of cotton in the Southern States was confined to a small " patch" on each farm, capable of producing a few pounds only, from which the seeds were picked by hand, and the wool washed, carded, spun, and woven by the women of the family into cloth or " homespun," then the chief article of clothing.

The staple production of South Carolina and Georgia, at that time, was indigo, and cotton was known, as an article of commerce, only in India and the East.

But, with Whitney's gin, which in a few minutes could pick the seed out from as much wool as a whole family could pick in a day, the sagacious people of these States perceived that the cultivation of cotton would be much more profitable than indigo—so they gave up indigo and undertook cotton. The farmers in India, perceiving how much superior to theirs the American cotton was, gave up cotton and undertook indigo—for the indigo of India is as good as ours.

About seventy years ago, an American ship, from Charleston, arrived in England with ten bales of cotton as a part of her cargo. She was seized, on the ground that so much cotton could not be produced in the United States. Before the war, the production had reached four millions and upward.

The invention of the spinning-jenny and the power-loom, about that time, tended still further to stimulate the production of cotton ; and as raiment is to the human family next in importance to food, the production of cotton in those States continued to increase until the year before the war, when it had reached the enormous quantity, before stated, of four millions of bales and upward, and which, at present prices, would be worth not less than $200,000,000.

8. Value of Productions.—Before the war, the people of the Southern States addressed themselves with great skill and energy to the various branches of agricultural industry, wisely trusting to the natural advantages afforded by their soil and climate to give effect to their labor ; and though numbering but little more than one-third of the population of the whole country, they produced, in value, two-thirds of the whole amount of its exports.

Questions.—1. Name the eleven Southern States.—Between what parallels of latitude do they lie?—What countries remarkable for fertility in the Old World lie between the same parallels? *2.* What is said of domestic servitude? *3.* How do the Southern States compare, as to climate, area, and population, with the New England and Middle States?—Contrast their coast-lines and compare their harbors, from Norfolk to the mouth of the Rio Grande. —What obstructs the harbors?—Where is Norfolk?—Trace out on the map, and describe, the chain of islands in the Atlantic that skirts the Southern coast. —How much water do the largest ships draw?—Describe the three principal watersheds into which the Southern States are divided, and point out the principal streams which carry off the drainage.—What parts of the Southern States suffer most from want of water?—Point out and describe the Llano Estacado. *5.* Point out on the map those parts of Virginia that are richest in minerals.—Those in Tennessee.—Those in North Carolina.—What kind of minerals?—What kind in Louisiana?—What kind in Arkansas?—In Alabama? *6.* How does the climate of Virginia compare with the climate of New England, on the one hand, and of the Gulf States, on the other?—What are the chief industrial pursuits of Virginia?—What, in the more Southern States? *7.* Can you give instances in which industries of people have been changed or created by human inventions?—Before the invention of the cotton-gin, what were the chief staples of South Carolina and Georgia?—Why did indigo go to India and cotton come to Carolina and the South for cultivation ? —Why was an American ship, with a few bales of cotton on board, seventy

years ago, seized in Liverpool?—How many million of bales had the annual
cotton crop reached before the war? 8. What portion of the exports of the
country at that time was of Southern growth ?

LESSON XXX.

The Southern States—Continued.

VIRGINIA.

Virginia, the oldest of the States, and "Mother of
Statesmen," was the largest of the "original thirteen,"
and used to be called the "Old Dominion."

In the Revolution of 1776 she took the lead, and
played a most conspicuous part. She was renowned for
the virtue of her sons and the wisdom of her statesmen.
Some of the greatest men—Washington, Madison, Marshall, Jefferson—that the country has produced, were
Virginians.

This State is situated between the parallels of 36° 30′
and 39° 40′ north latitude. It fronts for more than two
hundred miles on the Atlantic Ocean, and on the magnificent Chesapeake Bay, itself an arm of the sea. Its
western borders extend back to the tributaries of the
Ohio, and form a part of the Mississippi Valley. They
are drained into that river through the Ohio by New
river and the Tennessee.

The climates of Virginia correspond nearly with those
of Cashmere and the best parts of China.

Her latitude, the length of her days and nights, and the skies overhead, are
the same as those in some parts of Asia Minor; but the climates of the two
countries differ chiefly in this—Asia Minor is a dry country, Virginia is well
watered.

The mountains here, though they rise into peaks 4,000
or 5,000 feet high, are neither snow-capped nor barren,
but are clothed with forest-trees and undergrowth from
the bottom to the top, affording fine range and pasture
for cattle.

This woody vesture is a striking and peculiar feature of the whole Alleghany range. Both these mountains and their spurs are forest-clad from
Maine to Georgia. Upon them, as well as in the valleys between their ridges
and spurs, are to be found medicinal plants, timber, and ornamental woods of
various kinds and fine quality, such as cypress and cedar, maple, walnut,
chestnut, beech, wild cherry, dogwood, and lignum vitæ, pines and oaks
of many varieties, with hickory, ash, mulberry, snake-root, ginseng, sumac, etc

In tide-water Virginia, the cutting of ship-timber for
northern builders, and of fire-wood for northern brick-
kilns and other purposes, creates profitable industries.

The Natural Bridge, in Rockbridge County, is an
object of great interest to tourists.

NATURAL BRIDGE.

The Alleghany Mountains, and their outlying range
of the Blue Ridge, run along in a zigzag course, but
nearly parallel to each other ; the valleys between them
vary in breadth from 30 to 70 miles, and are very fertile.
The Valley of the Shenandoah is the largest and most
fertile among them.

Rising on the eastern slopes of these mountains, and
flowing through these valleys, are the Potomac, the
Shenandoah, the James, and the Roanoke rivers. The
Rappahannock and the York rise east of the Blue Ridge.
The New river has its head-waters in North Carolina,
and empties into the Ohio river, on the western slopes
of the mountains, where the head-waters of the Tennessee
take their rise. These are noble rivers. The smallest
of them is larger than the Thames in Europe.

All these rivers which flow down the Atlantic slope, are navigable from
the sea to the head of tide-water, and for distances varying from 100 to 200
miles from the ocean.

The distances from their sources to tide-water varies from 50 to 250 miles,
with a total fall in their descent of from 300 to 3,000 feet in the aggregate.

This affords, all along these streams, from the mountains to tide-water,
abundant water-power for mills and machinery of all sorts.

The whole country is well wooded and watered, and is
rich in minerals.

The coal-fields near Richmond have been profitably worked for many
years. The coal is bituminous, and is extensively used in Philadelphia, New
York, and other cities, for the production of gas.

Winchester, in the Shenandoah Valley, and Fred-

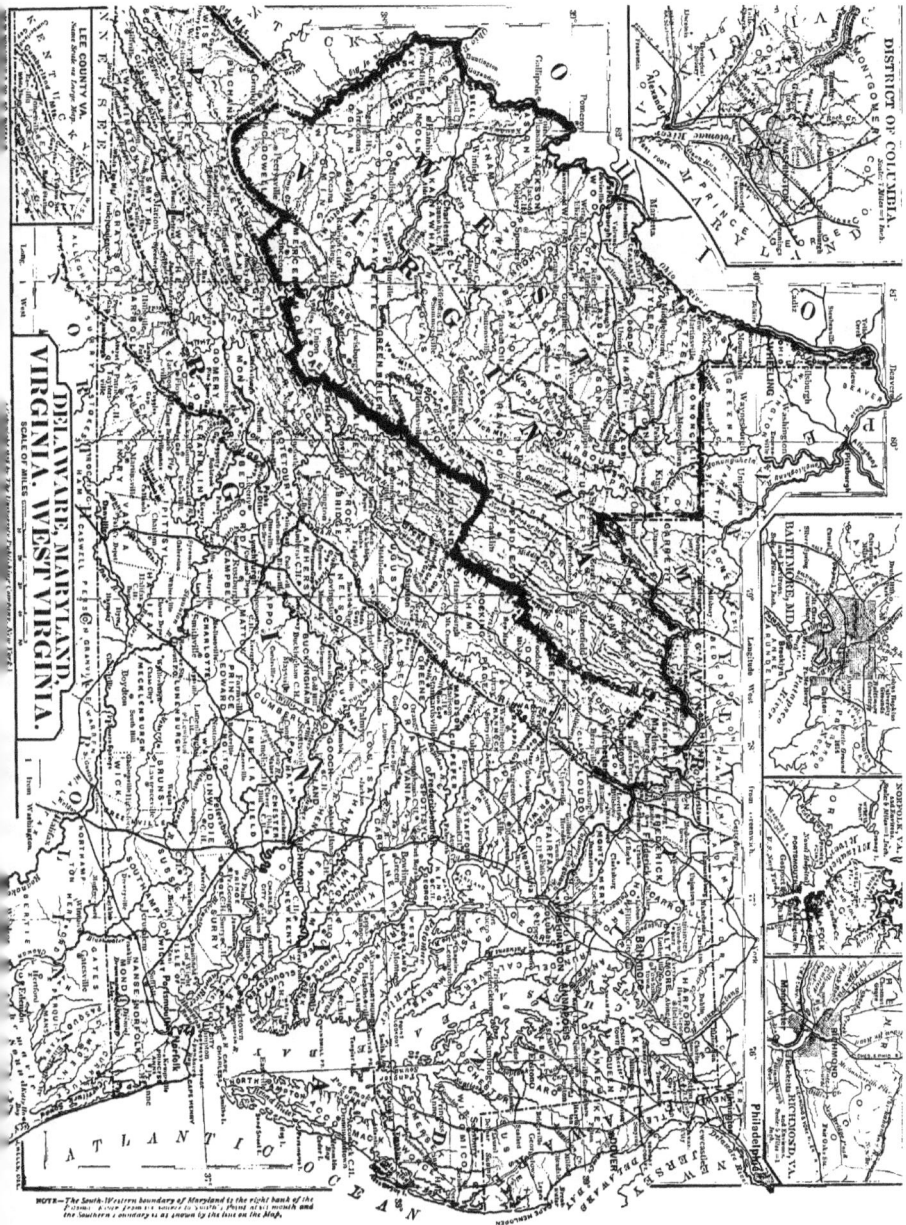

DELAWARE, MARYLAND, VIRGINIA, WEST VIRGINIA.

SCALE OF MILES

DISTRICT OF COLUMBIA.

BALTIMORE, MD.

NORFOLK, VA.

RICHMOND, VA.

LEE COUNTY VA.

NOTE.—The South-Western boundary of Maryland is the right bank of the Potomac river from its source to Smith's Point at its mouth and the Southern boundary is as shown by the lines on the Map.

TENNESSEE
SCALE OF MILES.

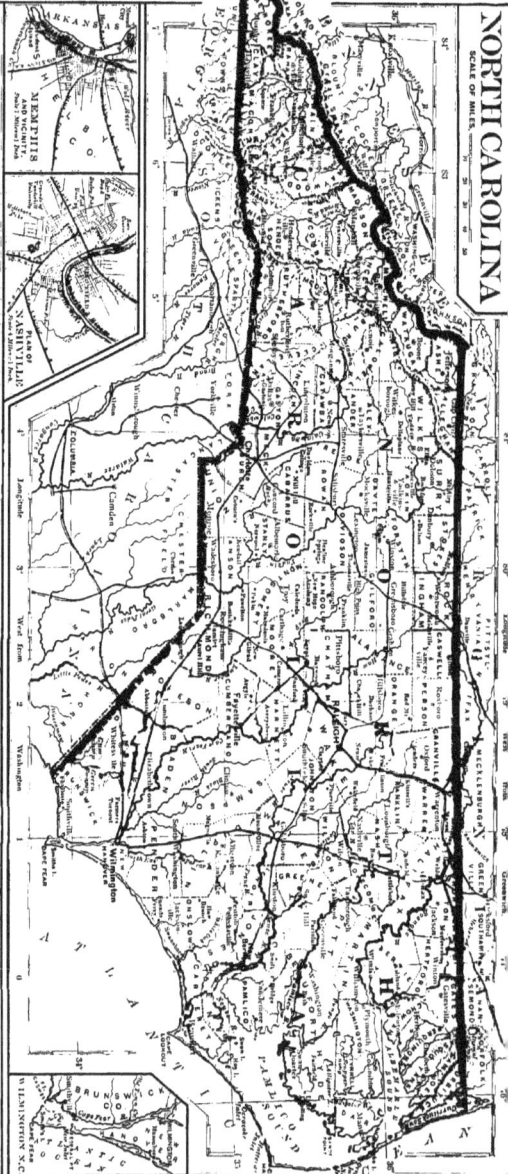

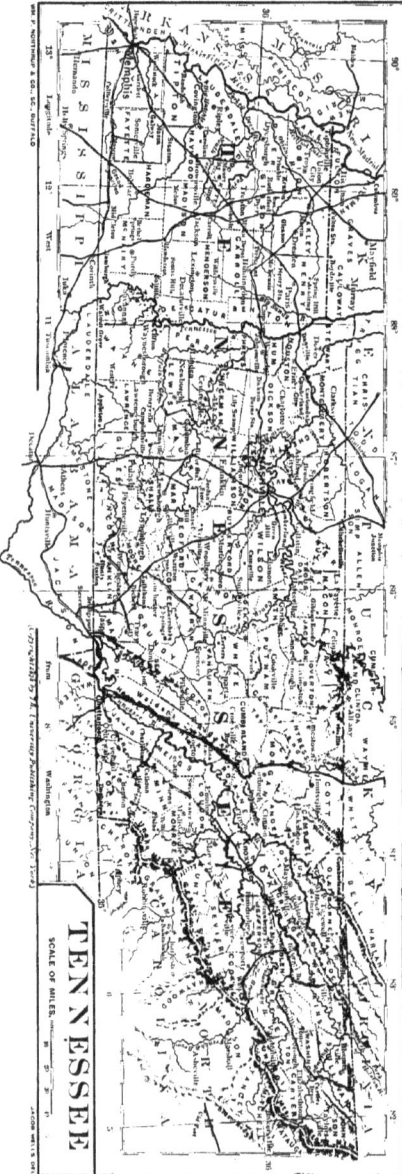

MEMPHIS AND VICINITY.

PLAN OF NASHVILLE.

ericksburg, on the Rappahannock, are celebrated for the great battles that were fought at and near them during the late war. The latter has fine water-power. Williamsburg was, in colonial times, the capital of the State ; and William and Mary, situated there, is the oldest college in Virginia. The University at Charlottesville and the Virginia Military Institute with Washington College, both at Lexington, are flourishing institutions.

The harbor of Norfolk, for capacity and depth of water, is not surpassed in the United States.

Richmond has most extensive flour-mills, large foundries, and a great number of machine-shops.

Richmond flour is especially valued, because, in shipping it across the Equator, it is not liable to "heat" or ferment.

Petersburg and Lynchburg are largely engaged in the manufacture of tobacco, for which the climate of the latter is particularly favorable.

Staunton, in the valley of Virginia, has asylums for the blind, the deaf and dumb, and for the insane. Alexandria, eight miles below Washington, has a good trade.

NORTH CAROLINA AND TENNESSEE.

North Carolina and Tennessee are between the same parallels of latitude, and, except in the tide-water country of the former, the industrial pursuits of the two States, so far as the soil is concerned, are very much the same. Tennessee is the daughter of North Carolina, as Kentucky was of Virginia.

The . territory once belonged to her, and Tennessee was settled chiefly by emigrants from North Carolina. Daniel Boone, the celebrated backwoodsman, who led the way for settlers, both into Kentucky and Tennessee, was a North Carolinian.

The cypress swamps and forests of pitch-pine, which abound in the tide-water country of North Carolina, afford to her people important and valuable branches of industry, in the cutting

SCRAPING CRUDE TURPENTINE.

and getting out of cypress staves, shingles, of lumber and naval stores.

Tar, pitch, and turpentine are all the productions of the yellow pitch-pine. The turpentine is obtained by blazing the tree, and dipping the gum from a box that is put at the root to receive it as it exudes from the tree.

Many of these trees are very tall and straight, and they make the finest masts and spars for ships in the world ; large numbers of them are sent to the dock-yards of France and England for their men-of-war.

The mild climate and the tides in the flat country of North Carolina, adapt many parts of it to the cultivation of rice, and in the geographical distribution of labor in this State the people find profitable branches of industry in their rice-fields, as well as in their pine-forests, the presence of which, in Tennessee, is forbidden by geographical law. Both States are admirably adapted to the growth of Indian corn, wheat, rye, oats, peas, beans, and barley ; flax and hemp ; to the vine, fig, and peach, with other orchard-fruits ; to melons, peanuts, and sweet potatoes ; and along the southern borders of both States cotton is extensively cultivated.

In the mountainous portions of these States are found valuable deposits, and veins of gold, copper, tin, lead, iron, coal, and marble. The marbles of Tennessee are more esteemed than those of any other State for their beauty and variety. Tennessee also excels in stock-raising.

Mount Mitchell, the highest peak of the Alleghanies, is in North Carolina.

The University of North Carolina, at Chapel Hill, is an old and excellent institution.

Wilmington is the chief place of export for the naval stores, staves, shingles, timber, rice, and cotton of North Carolina. It has many saw-mills.

Newbern is famed for its Indian corn, peanut, sweet potato, and melon trade with the North ; Albemarle Sound for its fisheries,—more than a million of herrings are sometimes caught there at a single haul.

Memphis and Chattanooga, in Tennessee, have not been surpassed in the rapidity of their rise as places of importance by any towns of their size in the South. The former derives its importance from its situation on the left bank of the Mississippi; the latter, from its situation at the junction and crossing of a grand system of railroads. Memphis is the chief cotton port for the planters of North Alabama and Mississippi, who send their crops there to be shipped by steamboat to New Orleans. Norfolk has of late become the chief shipping port of Tennessee ; for she sends, by rail, more cotton to Norfolk for shipment thence by sea, than she sends to New Orleans *viâ* Memphis and the Mississippi river.

The Tennessee river is navigable from its mouth up to Florence, Alabama, at the foot of Muscle Shoals.

7

These celebrated rapids—twelve or fifteen miles long—are formed by the river as its waters rush over that remarkable vein of limestone which gives us elsewhere the Falls of Niagara.

Attempts have been made to improve the navigation of these rapids by a canal like that round the falls of the Ohio at Louisville. United States engineers are now at work upon them. These officers, coming from the severer climates of the North, are charmed with the lovely climates and fine country here.

This river drains 15,000 square miles of country above the Muscle Shoals, with 825 miles of natural navigation, which is capable, with inexpensive improvements, of being extended to 1300 miles.

The advantages of this section of the country over the Northwest must have their weight; and when it is more generally known that its climate permits the Malaga grape, the fig, and the pomegranate to flourish in the open air, immigration must be turned to the Tennessee valley in the vicinity of Chattanooga and Huntsville.

The vast resources of this lovely valley, as an agricultural and stock-growing district, are demonstrated by the fact that upon them both armies subsisted for nearly two years during the late war.

The coal deposits of Hamilton and Roan counties, in this State, are enormous, and the coal is of a quality equal to the best Pittsburg coal for all purposes.

SOUTH CAROLINA AND GEORGIA.

South Carolina and Georgia resemble each other in climate. They both front on the Atlantic and abut against the mountains. Their industries are the same, and we speak of them together.

South Carolina takes the lead in the production of rice. Of all the Southern States, she and Georgia were the foremost with railroads.

The University of South Carolina is in Columbia, a beautiful country town and an elegant capital.

Charleston is the principal city and chief seaport town of the State, but like all Southern ports, the entrance to it, for large ships, is obstructed by a bar.

The palmetto grows in the streets of Charleston. As an emblem of sovereignty it was borne on the shield of the State; for that reason she is called the Palmetto State. The palmetto is a tree-palm.

Of all trees, those of this family are the most useful and beautiful. Among its varieties—of which there are not less than 60 in the "New World," which are entirely unknown in the "Old"—are found specimens which furnish man with food and shelter, with weapons and garments.

The shores of Georgia and the south coast of Carolina are curtained with the "Sea Islands," which are celebrated for a superior kind of cotton, called the "Sea Island" cotton.

This cotton, formerly growing only in these Islands, is now also cultivated very successfully in Southeastern Texas and Southern Louisiana, near the Gulf Coast. It has a long silky fibre, and is chiefly used in Europe, especially in Brussels, for the manufacture of laces and other fine fabrics. It is sold at four or five times the ordinary value of other cotton.

Rice is one of the chief staples of both South Carolina and Georgia.

In Georgia the seasons are so far in advance even of those no farther north than Virginia, that it is no unusual thing to see green peas and strawberries grown in the open air and fit for table use, in March.

The State University is at Athens. Augusta and Atlanta are celebrated for their workshops. Georgia is more extensively engaged in manufacturing than any other Southern State.

RICE PLANT.

The gold mines of Georgia, as well as those of North Carolina and Virginia, have been worked with profit, and before the gold mines of California had revealed their richer treasures, they were considered very rich.

The climates of Georgia, on account of its low latitude on one hand and its high mountains on the other, are very varied.

The hill country of Georgia, like that of Tennessee, produces the finest of wheat, while the rice delights in the low country along the coast.

Savannah is the chief city of Georgia.

BAY VIEW OF SAVANNAH.

FLORIDA.

Florida has the mildest climate of all these States. It fronts both on the Gulf of Mexico and on the Atlantic, and though its winters are too warm for frost, its sum-

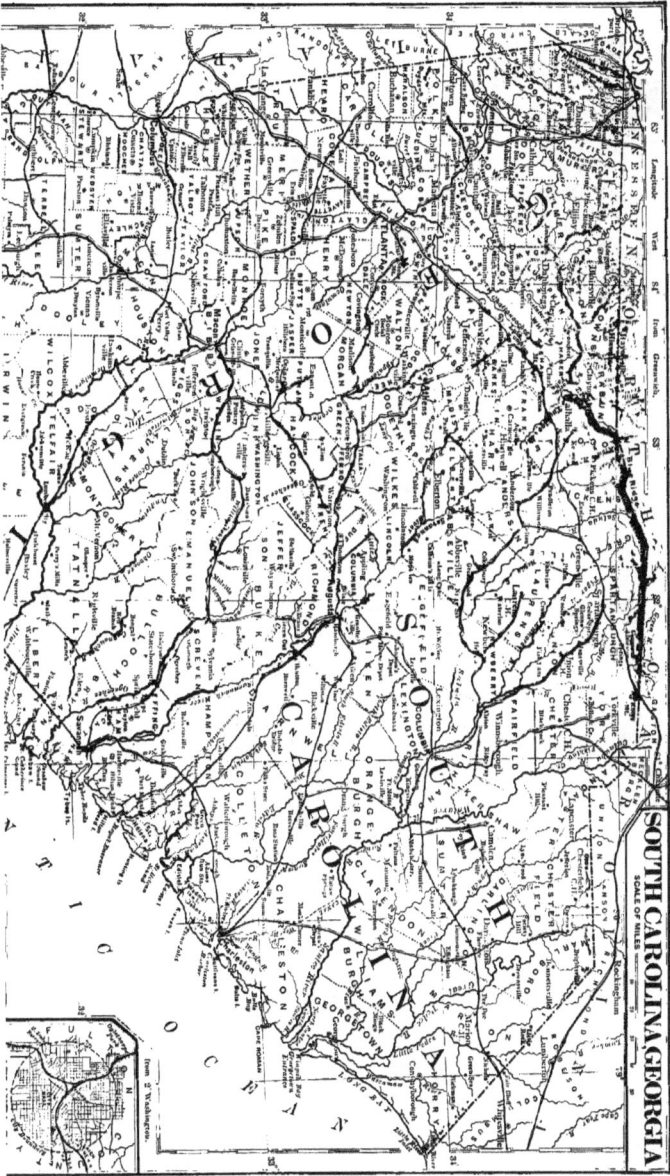

SOUTH CAROLINA AND GEORGIA

SCALE OF MILES

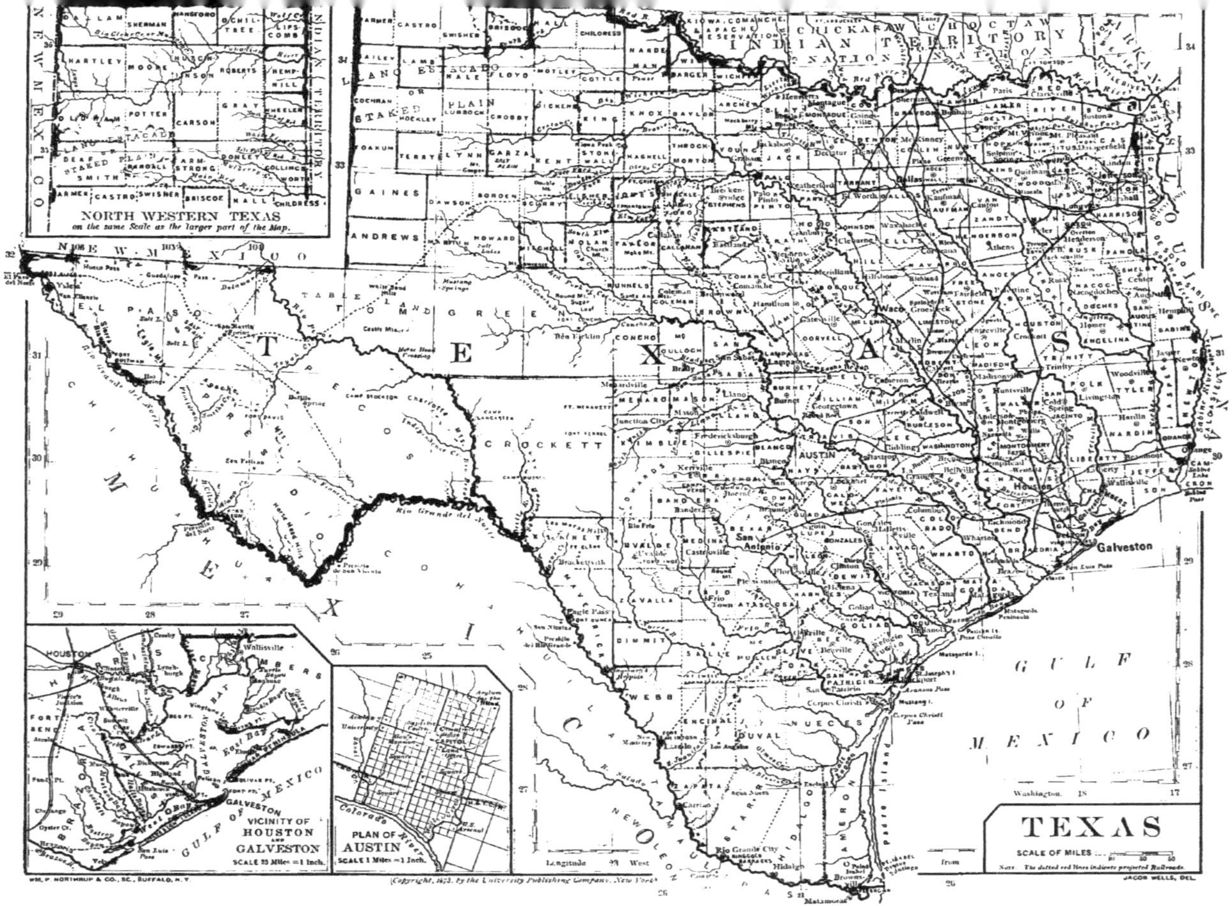

TEXAS

SCALE OF MILES

NOTE. The dotted red lines indicate projected Railroads.

JACOB WELLS, ENG.

NORTH WESTERN TEXAS
on the same Scale as the larger part of the Map.

VICINITY OF
HOUSTON
AND
GALVESTON
SCALE 17 Miles = 1 Inch.

PLAN OF
AUSTIN
SCALE 1 Mile = 2 Inch.

GULF OF MEXICO

NEW MEXICO

INDIAN TERRITORY

CHOCTAW
CHICKASAW INDIAN TERRITORY

Galveston

Padre Island

mers are so tempered by the sea-breezes and the ocean that the heat is less oppressive there than it is in New York and other States.

Though the Spaniards established a settlement at San Augustine, in Florida, long before any other Europeans had begun to found colonies in America, this State is so thinly settled at this day that there is only one inhabitant for every 300 acres of land; in New York there is one for every 6¼ acres.

Florida is adapted to the cultivation of all that is grown in the four other Atlantic States immediately to the north of it, with the addition of the sugar-cane, of intertropical fruits, and even of coffee in favored spots.

Florida is famous for its oranges and other fruits. The sweet potato produces there until killed by the frost, and in the southern parts the people gather it from the same patch, without replanting, for two or three years consecutively, and until there comes a killing frost, so mild is the climate. This State is mountainless, but abounds in swamps and everglades, and its live-oak forests are one of its chief ornaments.

Live-oak is the hardest, the heaviest, and most durable of woods; it is considered well-nigh impervious to decay, and is, on that account and for its strength, extensively used in ship-building.

The soil of the country is of limestone and coral formation. It abounds in beautiful lakes and clear, deep springs.

Some of the latter, of lake-like proportions, are deep enough to float a line-of-battle ship, yet so limpid that the pebbles can be distinctly seen on the bottom.

The Gulf Stream sweeps around this State and separates it from the great Bahama banks and islands, which are also of coral—making navigation dangerous.

The *Dry Tortugas*, off the coast of Florida, belong to the United States, and are fortified.

Key West is a famous wrecking station, where the property rescued from shipwreck is brought to be disposed of. Pensacola has the deepest water of any harbor on the Gulf-coast of the United States. The Government has a navy-yard there.

Questions.—Between what parallels of latitude does Virginia lie?— What is the length of her coast-line?—What part of Virginia lies in the Mississippi Valley?—Through what rivers is it drained into the Mississippi river?—What parts of the Old World are in the latitudes of Virginia?— Which of them resemble her most in climate?—What is the chief difference? —Describe the mountains in Virginia.

Name some of their ornamental woods and medicinal plants.—What are the principal rivers in Virginia?—Which empty into the Chesapeake?— Where does New river rise and empty?—The Tennessee?

How far are the rivers, that are tributary to the Chesapeake, navigable?— What, from their source to the head of tide-water, is their total fall?—What, the distance?

What kind of coal is mined near Richmond?—Where is Staunton, and for what is it noted?—Fredericksburg?—Where is William and Mary College?—Where is the University of Virginia?—Where, Washington College, and the Virginia Military Institute?—Point out the places where they are, and tell their bearings from Richmond.—What is said of Norfolk harbor? —What can you say about Richmond?—Petersburg?—Lynchburg?— Staunton?

What State has the finest marble in the United States?—Where and what

is the highest peak of the Alleghany Mountains?—Where is the University of North Carolina?—What can you tell about Wilmington?—Newbern?— The fisheries of Albemarle Sound?—In what rural industries does Tennessee particularly excel?

To what do you ascribe the rapid rise and importance of Memphis and Chattanooga?—Does Tennessee export most cotton viâ New Orleans or Norfolk?—To what town is the Tennessee river navigable from its mouth? —Where are the Muscle Shoals?

Upon what vein of rock are the Muscle Shoals?—How far is the river navigable above them?—What is the area of its valley above them?—What account do the officers of the engineer corps of the army give of this valley?— Of its climates?—Of its agricultural and mineral resources?—In what part of the State are the Tennessee and Cumberland rivers?

How do South Carolina and Georgia resemble each other?—In what branch of industry does South Carolina take the lead of Georgia?—In what line of improvements are these two States ahead of the other Southern States? —Where is the University of South Carolina?—Why is South Carolina called the "Palmetto State?"—Are there many varieties of the palm-tree?—How many are there in the New World?—Where does the Sea Island cotton grow?—What is it used for?

How, as compared with those in Virginia, are the seasons in Georgia?— Where is the State University of Georgia?—For what are Augusta and Atlanta noted?—Which of the Southern States is most extensively engaged in manufacturing?—What part of Georgia is best for wheat?—What for rice? —How does Florida compare in density of population with New York?— Is Florida thickly settled?—What can you cultivate in Florida that cannot profitably be grown in Georgia and the States north?—What delicious fruit is abundantly cultivated in Florida?—Describe the face of the country.—The springs.—What excellent ship-timber abounds there?—How is Florida separated from the Bahama Islands?—What makes the navigation along Florida coast so dangerous?—Where are the Dry Tortugas?—Where is Key West?— For what is it noted?

LESSON XXXI.

Southern States—Continued.

ALABAMA AND MISSISSIPPI.

With the exception of the hilly regions in the north-east corner of Alabama, the face of the country in these two States is similar. Their latitude and climates are also much the same. Cotton is their staple production, in which they excel all the other States, as Alabama and Tennessee did, according to the census of 1860 in the production of corn.

The pine forests and cypress swamps of North Carolina extend all the way along the coast to the mouth of the Mississippi river, and even beyond.

The Mississippi river, as it flows through the lowlands of the South, is prevented, in many places, from overflowing its banks and converting these low grounds into swamps, by embankments called *levees*. In this way a vast extent of land, remarkable for its fertility, has been reclaimed.

These reclaimed lands were known as the "Mississippi bottoms."

Before the levees were constructed, the whole area of lands in the Mississippi Valley, subject to overflows, and therefore unsalable, was estimated to

SCENE ON THE MISSISSIPPI RIVER.

be not less than 34 millions of acres. The State of New York does not contain as much as 34 millions of acres.

The rain-fall in the Southern parts of Alabama, Mississippi, and Louisiana (60 inches) is nearly twice as heavy as it is between the same parallels of latitude in Georgia and Florida on the one side, and in Texas on the other. It is greater than in any part of the country east of the Rocky Mountains.

The University of Alabama is at Tuscaloosa, and that of Mississippi, at Oxford. The latter has been successfully revived since the war.

Alabama has rich deposits of coal and iron, but Mississippi lacks coal, and is poor in metallic ores of all sorts.

Mobile is one of the two gulf ports from which most of the cotton produced in these States, as well as in those of Louisiana and Arkansas, finds its way to the sea and to distant ports.

LOUISIANA.

Louisiana was settled by the Spaniards and French; the descendants there of the latter are called French

Creoles, a term that has been borrowed by the natives generally, who call themselves Creoles instead of native Louisianians.

Louisiana was purchased from France in 1803, chiefly to secure a free outlet to the sea; but, in the purchase, was included all the country west of the Mississippi, even as far as the Pacific Ocean, except California, Texas, New Mexico, and Arizona.

During the war, when the people of the South were suffering for the want of salt, the usual supplies of which, except in Virginia, being cut off, an island of excellent rock salt was discovered on the coast of Louisiana. It continues to be profitably and extensively mined.

New Orleans is the great emporium and produce market of the South and West. It exports more cotton than any other seaport town in the world. At certain seasons of the year its levees are piled up with produce that has been sent there from the up-country for exportation, and its wharves are lined for miles with steamboats, shipping, and other craft that are engaged in the carrying trade.

NEW ORLEANS.

The waters of the Mississippi are very smooth; the tall trees and thick forests on the banks break the violence of the winds, therefore the Mississippi steamboats are built to stand high out of the water, Some of them are like floating palaces, they are so large and splendid.

All along the banks of this river, and those of its navigable tributaries, the business of cutting and hauling wood for these steamers (for wood is their favorite fuel) is an important branch of industry. It was worth, before the war, seven or eight millions of dollars annually. The industry and energies of the people of this State are directed chiefly to the cultivation of cotton and sugar. The climates along the Gulf-coast are semi-tropical, and many of the fruits and flowers of the torrid zone, such as the magnolia-grandiflora, the orange, pomegranate, and fig, flourish there in great beauty and perfection.

From the mouth of the Red river to the Gulf the level of the Mississippi and its outlets—called Bayous—is higher than that of the adjacent country. There the drainage is *from* and not *toward* the water-courses. There the people say, "Let us go *up* to the river," instead of "*down*" to it, as we do. In this low and flat country, called "the coast," the river banks are the highest lands. The palm-leaf fans that you use in summer come from these swamps and marshes, with their exuberant vegetation.

The depression of the country lying on both sides of the Mississippi, south of the Red river, exposes it to fearful floods and inundations.

In the lowlands and swamps, from North Carolina, extending along the seaboard, and for many miles back in the interior, all the way to Texas, the forest-trees of the South are draped in gray moss, a parasite that hangs down in long and graceful festoons from the branches, imparting to the forest scenery a striking and picturesque feature.

This is the moss which is so extensively used in upholstery for beds, cushions, and mattresses.

TEXAS.

Texas lies between the parallels of 26° and 36° 30'. It is the largest State, as to area, in the Union, though it has a population of less than seven persons for every square mile—Massachusetts, one hundred and eighty-five. Texas is thirty-five times the size of Massachusetts.

Texas was formerly a part of Mexico; she separated from that country in 1837; her independence was acknowledged, and then, in 1845, she was annexed to the United States.

In Northern Texas the atmosphere is dry, and the quantity of rain which falls there annually is small.

A dry climate makes hot summers and cold winters, a fact which will be explained in the Physical Geography, but which should be remembered, for it will serve you as a key to the climates of many countries.

In some parts of Texas the climate is admirably adapted to the cultivation of cotton and sugar ; in others, to corn, the olive, and the vine. Texas is also a fine grazing and wool-growing country.

San Antonio is the oldest town in Texas. Houston is a flourishing shipping port for a large section of rich country. Galveston is the chief port of Texas, as it has a fine harbor.

The dry part of the State begins with the celebrated "Llano Estacado" (the staked plain) already spoken of, that borders Texas and New Mexico. It is about 200 miles from east to west, and 300 from north to south.

Texas is famed for its beautiful prairies and

the severity of its north winds. These come on at times so suddenly in winter, and are so cold and severe, that both man and beast have been known to perish in them.

There are, in Texas, New Mexico, and also in the Indian Territory, vast plains which abound in prairie dogs, buffalo, wild deer, and other game.

ARKANSAS.

Arkansas abounds in swamps and lowlands. About one-fourth of the State was liable to overflow before the system of *leveeing* the Mississippi was commenced. The famous Red river raft, which was so instructive to the geologists of Europe, is in Louisiana, near this State.

There the driftwood has lodged for ages. In that warm climate plants, vines, and creepers soon began to take root upon this mass of trees and logs which covered the river from one side to the other. Presently trees began to grow upon it : these, with their roots, tendrils, and branches, bound this drift matter in one compact mass. It extends miles up the river, which disappears from view as it flows under *the raft*, near Shreveport.

Arkansas has but few towns, and none of them are large ; thus indicating that her industry is rural.

The western part of the State is a good grass country, and, among its mineral resources, it has a quarry of the most valuable whetstones known to commerce. Valuable deposits of zinc, coal, iron, lead, and antimony, with perhaps copper and silver, are also known to exist within its boundaries.

The hot springs of Arkansas are celebrated for their medicinal virtues.

NEW MEXICO AND INDIAN TERRITORY.

These territories are both bounded on the north by the same parallel, of 37°. Congress set apart this In-

PRAIRIE DOG VILLAGE.

dian territory, and gave it to the red-men and their descendants, to be occupied and governed forever in their own way. The Cherokees, Chickasaws (*chick'a-saws*), Chocktaws, Creeks, and Seminoles, are the most noted of these tribes. They till the soil, and have a constitutional government, schools, and churches.

Tahlequah (*tah'le-kwah*), the chief town, is in the Cherokee division. Some members of this tribe have elegantly furnished houses, are accomplished in manner, and refined in taste. Some of their neighbors, however, in the old-fashioned way of their fathers, still scour the plains in search of game and adventure.

SCENE IN NEW MEXICO.

New Mexico was formerly a part of Mexico, and was settled by emigrants from that country. Spanish is still the language in most common use.

Its landscapes abound in grand and imposing scenery, but they too frequently lack the charm of green pastures and still waters. The face of the country is often as dreary and wild as naked rocks and barren wastes can make it; but, wherever there is water, the soil is, as it is in the Indian Territory, exceedingly fertile.

Certain varieties of potato thrive in New Mexico. Indian corn, wheat, and the small grains do well. Onions, cymlings, and melons attain to an enormous size,

and great excellence in flavor. The grape commences to ripen in July and ceases in October.

The celebrated "El Paso" wine—a superior kind of Madeira—is made from this grape.

The olive and the date would do well in that country.

Near Las Vegas, in New Mexico, are some celebrated hot springs. There is a cluster of thirty or forty of them, of various temperatures, from 80° to 140°.

Near the city of Colorado are four remarkable soda-springs, which, within a short distance of each other, come bubbling and boiling out of the earth as though they were fresh from the fountain.

Zuni is a small Indian village, situated in a desolate region.

There are some remarkable ruins in its vicinity, supposed to be the habitations of a former generation, made desolate by famine or pestilence.

Questions.—Which are the five Gulf States ?—Which has the most hill country, Alabama or Mississippi ?—Has either of them any mountains ?—How are the climates ?—What are their staple productions ?—In what branch of industry do these States excel all others ?—How far do the pitch-pine forests extend ?—Can you give an example of the effect that soil and climate have upon the industries of people ?—How, since the Mississippi river is higher than the country a little way back, is it prevented from overflowing its banks at high water, and drowning these low grounds ?—What are the "bottoms ?"—levees ?—What, before the levees were built, was the area of land in the Mississippi valley that was subject to overflow ?—What part of the country on this side of the Rocky Mountains receives the heaviest rain-fall ?—What is the depth of this fall in inches ?

What State has the largest city in the South ?—What city, of all in the country, exports the most cotton ?—How far from New Orleans to the Gulf? (See map.)—Describe the appearance of the city.—Why are the Mississippi steamboats built to stand high out of the water ?—To what important branch of industry have the steamboats of the West given rise ?—What are the two chief branches of industry in Louisiana ?—Which way does the cotton go for a market ? *Ans.*—Largely abroad.—How are the climates of Louisiana ?—What tropical fruits and flowers flourish there ?—Describe the lowland forests along the seaboard from North Carolina to Texas.—What use is made of this parasite ?—By whom was Louisiana settled ?—When and from whom was Louisiana purchased ?—What was the main object of this purchase ?—What extent of territory was included in this purchase ?—What valuable mineral deposit was recently discovered in Louisiana?

Between what parallels of latitude does Texas lie ?—(See Map, p. 57.)—How does it compare with the other States as to area ?—What population does it average to the square mile ?—How many times larger than Massachusetts is Texas ?—Suppose it were as thickly inhabited, how large would its population be ?—When was Texas annexed ?—Which is the driest part of Texas ?—What effect has a dry atmosphere on the summer and winter temperature of a country ?—What is the chief port of this State ?—Whence are the rivers that flow through the dry parts of Texas fed ?—How large is Llano Estacado (staked plain) ?—How high is it above the sea-level? (See map.)—What wind is particularly severe in Texas ?—Are there any plains in Texas ?—What animals do you find upon them ?

How much of Arkansas is liable to overflow ?—Describe the Red River raft. —What is the industry of the State ?—Its minerals ?—Its springs ?

How are New Mexico and Indian Territory situated as to latitude ?

By whom, and for what purpose, was Indian Territory set apart ?—Describe the Indian settlements.—Chief town.—Habits of Indians.

To what power did New Mexico once belong ?—What language is in use ? —Its scenery ?—Soil and products ?

What is the "El Paso" wine ?—What is said of Zuni ?

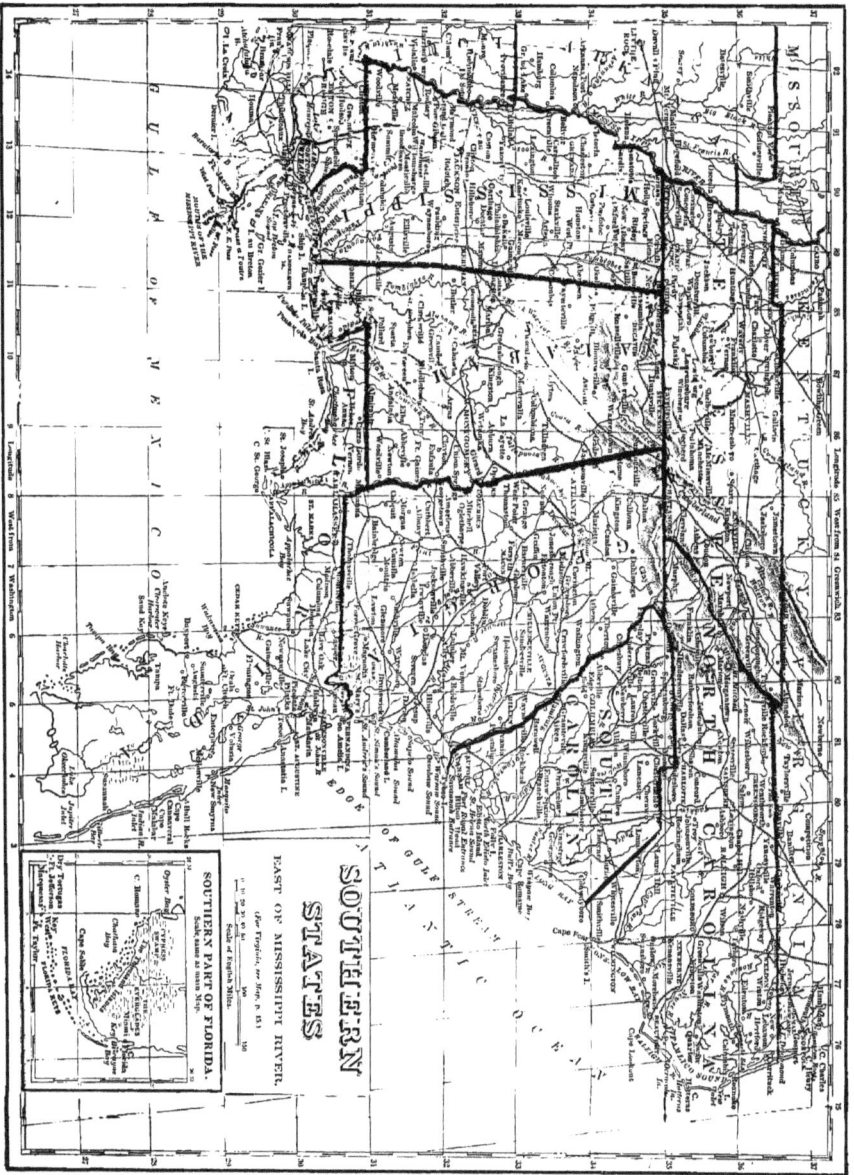

SOUTHERN STATES

EAST OF MISSISSIPPI RIVER.

(For Virginia, see Map, p. 51.)

Scale of English Miles.

SOUTHERN PART OF FLORIDA.
Scale same as main Map.

LESSON XXXVII.

Studies on the Maps of the Southern States and Territories. (Maps, pp. 32, 34, 45, 55, 57.)

Boundaries.—Bound the Southern States.—Bound each one separately.—Tell their capes.—Tell their bays and sounds.

Within what parallels of latitude and meridians of longitude are these States included? (For map of Virginia, see p. 45.)

Population, Capitals, and Chief Cities.—Tell their population.—Tell the situation of the capital and chief town, and the population of the chief town of each State.—Chief town of Indian Territory?—Of New Mexico?—Bound each of these territories.—Which is the smallest of the Southern States? Which is the largest?—On what river is Richmond situated?—Lynchburg?—Petersburg?—Fredericksburg?—Where is Staunton?—Winchester?—Charlottesville?—Salem?—Danville?—Point out Weldon.—Wilmington.—On what river is Columbia?—Montgomery?—Savannah?—Knoxville?—Nashville?—Memphis?—Yazoo City?—Vicksburg?—Natchitoches?—Monroe, La.?—Pine Bluff?—Batesville, Arkansas?—Austin, Texas?—Corsicana?—El Paso?—Point out Taliequah?—Santa Fé.—Albuquerque (*ähl-boo-ker'ko*).—On what river is Albuquerque?—Natchez above or below Vicksburg?

Routes and Directions.—How would you go to Richmond from Norfolk?—How, from Washington?—How would you reach Wilmington by sea from New York?—How would you go from Staunton to Richmond, in Virginia?—From Alexandria to Abingdon?—From Fredericksburg to the White Sulphur Springs?—From Danville to Fredericksburg?—From Richmond to Weldon, North Carolina?—From Danville, Va., to Raleigh?—From Fayetteville to Wilmington?—From Raleigh to Charleston?—From Charleston to Columbia?—From Columbia to Greenville?

Which way does Charleston lie from Richmond?—Which way from Wilmington?—How far, and which way is it, from Wilmington to Savannah?—From Charleston to Savannah?—To Augusta?—To Chattanooga?—To Knoxville?—To Memphis?—How far, and in what direction, is Eutaw from Savannah?—Tallahassee from Savannah?—How far from Mobile to Nashville?—From Mobile to Jackson?—How would you go from Mobile to Montgomery?—From Jackson to New Orleans?—From New Orleans to Shreveport?—From Shreveport to Marshall, Texas?—From Galveston to New Orleans?—From New Orleans to Little Rock?—From Yazoo City, in Mississippi, to Alexandria, Louisiana?—How would you go from New Orleans to Santa Fé?—How would you go all the way on water from New York to St. Louis?

Rivers.—What fine river separates Virginia and Maryland? How long is it, measuring by the scale of miles?—Name the other rivers of Virginia.—What is the long neck of land between the Rappahannock and the Potomac called? *Ans.* The Northern Neck.—What is that between the York and the James called? *Ans.* The Peninsula.—Part of Virginia is cut off by the Chesapeake Bay: its name? *Ans.* The Eastern Shore.—Name the chief rivers of North Carolina.—Of South Carolina?—Of Georgia.—Of Florida.—Where do most of these rivers rise?—How do most of them flow?—The principal rivers of Tennessee?—Their course, and sources?—Name the chief rivers of Alabama and Mississippi.—Where do they rise, and whither do they run?—Name the rivers of Louisiana and Arkansas.—Where do they rise?—What river do they unite with?—In what territory does the Arkansas river reserve tributaries?—Rivers of New Mexico?

Where is the Washita river?—Does it empty into the Mississippi or the Red?—Does the Arkansas empty above or below Memphis?—How far from Vicksburg?—What rivers form the Alabama?—Do they unite above or below the city of Montgomery?—Does the Red river empty into the Mississippi above or below the Mississippi State line?—What river partly separates Texas and Louisiana?

Lakes.—Name some of the lakes of Louisiana.—Are there any of them?—Name some of the lakes of Florida. Where is Lake Okeechobee? The Everglades?

Mountains.—Name the mountains of Virginia and North Carolina.—What is the chief peak?—Where is Lookout Mountain?—Find the southern spur of the Alleghanies.—Describe the mountains of New Mexico.—In what direction does the country slope?—How is its drainage carried off?

You observe that New Mexico, Colorado, and the Territories to the north, in which the Rocky Mountains lie, are at the top of the divide which separates the waters of the Atlantic from those of the Pacific.

What large rivers rise there, the one emptying into the Pacific, and the other into the Gulf of Mexico?

Capes and Keys.—Name the capes of Virginia.—Of Florida.—Point them all out on the map.—Point out Cedar Keys.—Key West City and Island.—Point out Cape St. Blas.

Islands and Inlets.—Where is Ocracoke Inlet?—Mosquito Inlet?—What islands belong to Virginia?—What, to North Carolina?—To South Carolina?—To Florida?—To Alabama?—To Mississippi?—To Louisiana?—To Texas?

Miscellaneous.—How far, by the map scale, is it from Florida to the Bahama Islands?—How far from Key West to Cuba?—Name bays and inlets of Florida.—Which of all the States has the greatest length of coast-line?—Where, and how wide, are the Straits of Florida?—Where is the Dismal Swamp?—Where is Port Royal Entrance?—Where is Tampa Bay?—Where is Cumberland Gap?—Where are Dry Tortugas Islands?—What great ocean current passes by Florida, and gives it a tropical climate?—In what parts of North Carolina are the Cypress Swamps?—The pine-forests?

SOUTHERN STATES AND TERRITORIES WEST OF MISSISSIPPI RIVER.

Scale of English Miles.

LESSON XXXIII.

The Western States and Territories.—Their Geographical Position and Features. (Maps, pp. 67, 73.)

TOTAL POPULATION, 14,813,713.

STATE.	Capitals.	Chief Cities and their Population.
West Virginia...........	Wheeling	Wheeling...... 19,280
Ohio	Columbus	Cincinnati...... 216,239
Kentucky...............	Frankfort	Louisville....... 100,753
Indiana	Indianapolis	Indianapolis..... 48,244
Illinois	Springfield..........	Chicago........ 298,977
Michigan	Lansing	Detroit......... 79,577
Wisconsin	Madison	Milwaukee...... 71,440
Missouri................	Jefferson City......	St. Louis 310,864
Iowa....................	Des Moines.........	Dubuque 18,434
Minnesota	St. Paul	St. Paul 20,030
Kansas..................	Topeka..............	Leavenworth ... 17,893
Nebraska................	Lincoln	Omaha 16,083
Colorado................	Denver..............	Denver........ 4,750
Montana Territory.......	Helena..............	Helena......... 3,106
Dakota "	Yankton	Yankton 737
Wyoming "	Cheyenne............	Cheyenne...... 1,450

1. Position and Orography.—These States and Territories are all inland, and all of them, excepting Michigan, give rise to streams that empty into the Mississippi river; they occupy what is often called the *Upper Mississippi Valley*. They also embrace portions of the two great watersheds, which are formed by the Rocky Mountains on the one hand and the Alleghanies on the other, so as to carry their waters off into the Mississippi, which lies between them as a gutter, and delivers the drainage into the Gulf of Mexico. This river, therefore, occupies the line of lowest level (see the Orographic View of the United States, p. 30) that can be drawn from north to south along the valley drained by it.

2. Course of the Mississippi.—You observe that the Mississippi does not occupy the middle of this valley, it is *far to the east of the middle*, winding along, especially in Kentucky and Tennessee, not very far from the out-lying ridge of the Alleghanies.

The geographical conclusions that, with the orographic view and the map of the United States before you, you are able to draw from this fact, are:—
1st.—That the eastern tributaries of the great river, as compared with the western, are more rapid in their descent.
2d.—That they are not navigable to so great a distance.
3d.—That the plains watered by them are not as broad or as long as are those through which the more gentle streams of the western watershed flow.
4th.—That when heavy rains occur, these long and gentle streams of the West require more time than do the shorter and more rapid streams of the East, to discharge their floods into the main gutter—the Mississippi river itself.

3. Climates.—The Orographic View shows all this, and more; it shows that these States and Territories, all lying between the parallels of 36° 30' and 49°, are separated from the sea by a range of mountains on the east and on the west; and you have already learned enough about the influences which regulate climates, to teach you that the differences of climate among these States and Territories are to be accounted for chiefly by mere difference of latitude and elevation, regardless of their distance from the sea.

You may also infer that those portions of these States that lie along the margins of the Mississippi and its great lakes are the warmest, because the lowest, while those parts that lie among the hills which give rise to its tributaries are the highest, and therefore the coldest, the latitude being the same.

4. Continental Slopes and Drains.—You may observe, by looking at the map of North America a little more closely, that from the Tropic of Cancer to the Arctic Ocean the continent is divided into two grand water-sheds, which, together, include all the minor ones that we have hitherto considered; and that, near the parallel of 50° north, and extending across our continent from the Pacific to the Atlantic Ocean, there is a ridge which divides the whole country into two grand watersheds, one of which inclines to the north, and drains the waters off into the Arctic Ocean and its Bays; the other inclines toward the south, and drains off into the Atlantic Ocean and the Gulf of Mexico, or into the South Sea, as the Pacific was called. The head-waters of these two drains—the Mississippi and the Red River—rise about 1,000 feet above the level of Lake Superior.

The distance between them is short, as it is between the Saskatchawan and Athabasca. So if there was a canal across these two narrow portages, an Indian entering the Mackenzie, from the Arctic Ocean, in his bark canoe, might, after ascending that river and passing these two canals, descend the Mississippi into the Gulf of Mexico, and so pass by fresh water channels from the regions of eternal winter to perpetual summer.

The ascent to the top of this ridge would be about 1,000 feet, and the descent on the other side from St. Louis to the Gulf being over 380 feet.

From Minnesota to the Atlantic this ridge seems to have been depressed and hollowed out, as it were, to make a *cistern* upon the top of it, and form a basin there for the great lakes. The St. Lawrence is the gutter for draining these lakes; it carries the water off to the east and empties it into the Atlantic Ocean.

5. Favored Position.—The regions occupied by the Western States are the granaries of the country. Every year they satisfy the land with bread; and after it is filled, they have enough to relieve famine abroad.

6. Population.—The thirteen Western States occupy an area of 767,000 square miles, and have a population of 19 persons to the square mile, or an average of 31 acres per inhabitant.

Belgium, which does not lie in such sunny climes, and whose soil is no more generous, averages only an acre and a half per head for her inhabitants.

According to this ratio, there is yet in these Western States, exclusive of the Territories, room for more than 290,000,000 of people.

The extent of their grassy plains and prairies, the cheapness of the lands, the facility with which they are brought under cultivation, together with the fertility of the soil, are the attractions which direct immigration to this part of the country in preference to any other portion of our wide domains.

7. Supply of Water.—The winter rains of the Pacific coast, to be treated of at another time, turn to snow on the Rocky Mountains, and there the snow remains as a reservoir to feed the rivers when it melts in spring and summer. Now, when the heavy rains of the spring and early summer happen to flood the eastern tribntaries at the time that the tributaries from the west are discharging great volumes, arising from a thaw in the mountains, the Mississippi river receives two floods at once.

The river sometimes swells over with these floods and attains the proportions of a sea. In the spring-flood of 1867, the Mississippi was estimated to be, at Memphis, more than 40 miles broad.

8. Prairies and Plains.—The most striking feature of the Western States is the size and picturesque loveliness of their treeless plains. They are covered with grass, gay with flowers, and alive with herds of wild cattle. On the east side of the Mississippi they are called prairies, and on the west side, plains. Nearly the whole of Indiana, Illinois, Wisconsin, and Minnesota is a level country, and the plains stretch out to the declivities of the Rocky Mountains.

9. Products.—The Western States are a grazing and agricultural country, and the staple productions are corn, wheat, oats, rye, and barley ; potatoes and culinary vegetables ; with hemp, grapes, fruits, and tobacco. The soil and climate are also admirably adapted to the growth of sorghum, or Chinese sugar-cane, which is rapidly assuming in its cultivation the proportions of an agricultural staple.

The people of the Western States are also extensively engaged in wool-growing, and in raising cattle, horses, beef, and hogs. The annual wool-clip both in Ohio and in Michigan is very great.

10. Minerals.—The Western States and Territories are richer than the Southern States in minerals. The coal-fields of West Virginia, Ohio, Indiana, and Illinois, are the largest and the most bountifully stored in the world. They embrace an area of many thousand square miles.

There are in the United States four great coal-fields. Those of Pennsylvania can yield 50,000,000 tons per annum. In Maryland the seam which supplies our steamers with the best fuel is 14 feet thick and 50 miles long. The coal-fields of Missouri alone have coal enough to last the world 3,000 years, while it would take 100,000 years to exhaust those of Illinois at the present rate of consumption.

Masses of native copper, tons in weight, have been quarried out of the mines in the Lake Superior copper region.

Lead abounds in Illinois, Wisconsin, Iowa, and Missouri. In the last-named State there is a mountain of iron.

Salt springs and wells abound on the east side of the Mississippi : in the mountains of the west, gold and

silver. Michigan now produces nearly as much salt as New York.

The fisheries of the great lakes are a profitable source of industry, especially to Michigan.

The white-fish of the lakes, when salted and packed, are highly esteemed as an article of food. The Mississippi river also yields bounteously of its fresh fish to the people along its banks.

Questions.—*1.* Do any of the Western States border on a sea or lake ? *2.* What is the course of the Mississippi ?—Name some of the conclusions you draw from a study of its course. *3.* Why have some of the Western States a different climate from others ? *4.* Name and describe the two great slopes of North America.—Do you observe any depression or hollow on the Orographic View which lead you to conjecture that there is here a grand natural cistern ?

5. Is the region occupied by the Western States favored ? *6.* What is the ratio of area to their population ? *7.* Where do the floods of the Mississippi come from ?—How can you illustrate their magnitude ? *8.* What is said of the Prairies ?—What are the Plains ? *9.* Products and industries of the people ?—Which States grow the most wool ? *10.* What is said of the mineral resources of the Western States ?—Where does lead abound ?—Where is there a mountain of iron ?—What is said of Michigan fisheries ?

LESSON XXXIV.

Western State — Continued.

WEST VIRGINIA. (Map, p. 45.)

This State was formed out of the "Old Dominion" during the last war. It is the most mountainous of the Western States, and is classed with them, because it adjoins them, is in the same latitude, lies on the western waters, and is an interior State, as they are. Its hills abound in coal and iron : in its valleys are to be found salt-springs, petroleum-wells, and mineral-waters of great excellence,—among them the celebrated White Sulphur Springs. No part of the world can surpass the mountains of the two Virginias in the abundance, variety, and excellence of their mineral-waters.

OIL WELLS.

Some of the finest varieties of coal known to commerce, such as splint and cannel coal, are found in the valley of the Kanawha river, on which Charleston, the former capital, is situated. It is at the head of slack-water navigation

There also are the salt-works. The water is obtained by boring down through the rocks below the bed of the river. The Kanawha salt is extensively used by the meat packers of Cincinnati.

The chief article of food for sailors at sea is salt beef and pork. Most of that used by the navy of the United States and in our merchant ships, and much also of that which is used on board the ships and navies of Europe, is packed in Cincinnati, and cured with the salt which comes from Kanawha and from Pomeroy, Ohio. This salt has peculiar properties, which give a special value to the meats that are packed with it.

The hills which surround Wheeling also contain valuable deposits of bituminous coal. This, with its position near the head of navigation of the Ohio, and its connections by railroads with Baltimore on one hand and the Western States on the other, gives Wheeling great importance.

It is proposed to enlarge the James River Canal and make it a great national ship-canal, capable of passing large steamers and other vessels to and fro, between the Ohio and Chesapeake, so as to give, in war, a water route between the Atlantic seaboard and the West, entirely within our own borders.

KENTUCKY. (Map, p. 67.)

This State was a colony of Virginia. It was settled chiefly by Virginians.

It is separated from the Atlantic seaboard by a mountain barrier, across which the passage in those early times, even on horseback, was difficult, and rendered perilous by the Indians.

Kentucky was separated from the Gulf of Mexico by a long and tedious river navigation, and to get her produce to markets then, she had to ship it on flat-boats and drift down with the current. Arriving at last at New Orleans, the cargo was sold, and the boat broken up for firewood—for it could not be poled back against the current; and the crew were left to find their way home on foot through almost pathless forests, infested, too, frequently by hostile Indians.

Thus it took one year to grow a crop, another to carry it to market and return, so that these early settlers could produce for market only one crop in two years.

Such was the condition of the Western States about fifty years ago, when steamboat navigation first reached the Mississippi River, and the revolution that steam and the steamboat have made upon the industrial pursuits of the people of these States is, with its effects, the most remarkable feature in their political geography. It has turned the howling wilderness into smiling gardens.

In many parts of the State the soil is of surpassing fertility. Lexington, in the midst of the famous blue-grass region, is on the vein of limestone that forms Niagara Falls, Mammoth Cave of Ky., and the Muscle Shoals of Alabama. It makes Kentucky rich in cattle.

MAMMOTH CAVE.

Hemp and tobacco, with corn and the cereals, are the chief agricultural staples of Kentucky. It is also a fine fruit country.

Iron mines are profitably worked in the lower part of the State. Coal also is abundant.

Kentucky and Tennessee are the only States east of the Mississippi that give their drainage entirely to that river, and are wholly within its valley.

Louisville is at the Falls of the Ohio, where there is a canal capacious enough to pass the largest steamers that ply on those waters.

Ohio.

The Ohio river, with its connections, placed this State in water communication, at an early day, with all the commercial marts of the Mississippi Valley. Her geographical position on the lakes, and the early completion of the Erie Canal, gave her great advantages, which were increased soon after by the construction through this State of several canals between the Ohio river and the Lakes. These works at that time contributed powerfully to the prosperity and rapid settlement of Ohio. At a later day the railways of New York, Pennsylvania, and Maryland completed the connection with the Eastern and Middle States, and made Ohio the great thoroughfare of trade and travel between the Atlantic seaboard and the West.

This State was also once a part of the "Old Dominion," and many of the first settlers of Ohio were Virginians.

There are no mountains in this State; the country is

comparatively level ; there is no lack of limestone and other rocks, and so it was quite easy for the early settlers, before steam was introduced as a locomotive power, to interlace this State with good turnpike roads.

Ohio has already become the third State of the Union, and her people are more largely engaged in mining and manufacturing than those of any other Western State. She has a growing trade with the South. There are valuable deposits of coal and iron in the region round about Ironton.

Ohio is a fine grain country, and the grape is extensively cultivated there for the manufacture of wine. The Catawba of Cincinnati is classed in Europe with the favorite wines of the Rhine.

Besides Cincinnati, which is the largest city of Ohio, Cleveland, Toledo, Sandusky, and Columbus (the capital of the State), are important and flourishing cities.

Pork-packing is the branch of industry for which Cincinnati is most noted.

INDIANA AND ILLINOIS.

These States are in the Prairie country ; they have no mountains ; their latitude, climate, and agricultural staples are the same.

Next to Ohio, Illinois is the most populous of the Western States, and, like Ohio, is greatly favored in its geographical position.

Chicago is one of the most flourishing cities in the West. It is especially remarkable for its grain and provision trade. It is the chief place for the shipment of grain and breadstuffs from the West to the East. Excepting London, Chicago is the greatest grain market in the world. Vessels sometimes take in the cargo at Chicago, and sail thence direct for Europe. Galena is in the midst of the lead-mining regions. Cairo derives geographical importance from its position at the junction of the Ohio with the Mississippi, and that importance has been vastly increased by the Illinois Central Railway.

Evansville, Indiana, is a flourishing city.

New Albany is a boat-building place, where many of the steamers that ply on the Mississippi are launched.

A new and magnificent bridge now connects Louisville with New Albany.

Questions.—What kinds of coal are found in the Kanawha valley?—For what is the Kanawha salt especially valuable?—How is the salt water obtained there?—What kind of coal is found in the hill-sides at Wheeling?—What waters is the James River and Kanawha Canal to connect?—Why is it proposed to make it a national work?

To what State did the territory which now forms Kentucky formerly belong?—By whom was it settled?—What made the journey for the early settlers there so difficult and dangerous?—How, before the introduction of the steamboat on the Western waters, did Kentucky get her produce to market?—How long did it take them to make and get a crop to market?—What is the most remarkable feature in the political geography of the Western States?—Can you give another example of the bearings of human inventions upon the geography of a country?—Where is the blue-grass region of Kentucky?—What great natural curiosity is found there?—What are the chief agricultural staples and mineral productions of Kentucky?—What two States east of the Mississippi lie wholly in the valley of that river?—Where is Louisville?—How do steamboats pass the falls there?

What State was the first to send settlers to Ohio?—Describe the face of the country.—How came Ohio to be a thoroughfare between the West and the East?—What advantage do the farmers of Ohio now enjoy on account of her geographical position?—How, in the order of population, does Ohio rank among the States of the Union?—How, among the Western States?—Which one among the States of the Union is the first in population?—First in area?—Besides the agricultural, what are the industries that chiefly engage the attention of the people of Ohio?—In what part of the State are mines of coal and iron extensively worked?—What celebrated wine comes from Ohio?—For what business is Cincinnati so noted?

Why are the staple productions in Illinois and Indiana the same?—In what respect is Illinois so favored in position?—Which two are the largest of the Western

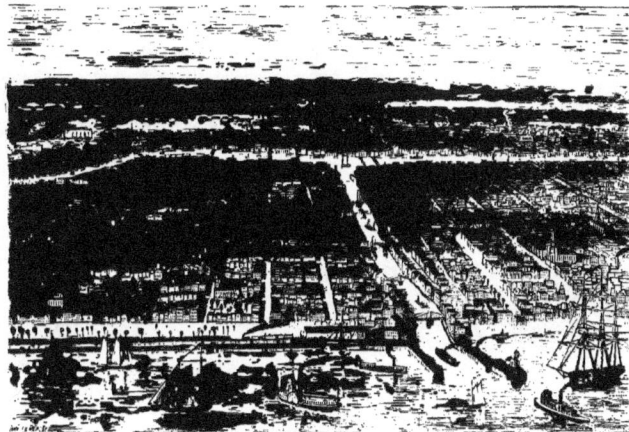

CHICAGO.

States?—Upon what rivers and lake does the State of Illinois border?—Upon what river and lake does the State of Indiana border?—Judging by the eye, which has the most river-front? (see Map.)—In what branch of business is Chicago especially noted?—Do vessels ever go from Chicago to Liverpool? —Where is Galena, and for what is it noted?—What increases the importance of the geographical position of Cairo?—For what branch of industry is New Albany noted?—Where is Evansville?

LESSON XXXV.

Western States—Continued.

MICHIGAN

Is divided into two peninsulas (*see Map*). It is nearly surrounded by the great lakes, and, like Ontario, its climate is milder than that of any of the other States in the same latitude. Its shores, that look out upon Lake Michigan to the west, are, on account of their softened climates, excellent for fruit culture.

Though navigation on the great lakes is annually obstructed by ice for about five months, they are frozen *entirely* over only for a short time.

Fresh water can never be colder than 32°, because that is its freezing point, when it becomes ice. In winter, when it is very cold, water at 32° feels comparatively warm, as you know by putting your hand into the water of a boiling spring. With extensive sheets of open water, like those on Lake Michigan in winter, to temper the biting west winds as they approach the eastern shores of the lake, you can imagine that the winter temperature of these shores is very much milder than that of the opposite shores.

PICTURED ROCKS OF LAKE SUPERIOR.

Lake Superior is the largest of the great lakes, and the highest above the sea-level, as is shown by the Falls of Sault St. Marie, over which its waters escape into the lakes below. A ship-canal has been constructed around the Falls, so that vessels may now pass to and fro between Lake Superior and the ocean. The Pictured Rocks of Lake Superior are often visited.

Lakes Michigan, Huron, and Erie are on the middle level, for the water is poured from them over the Falls of Niagara into Lake Ontario, which is still considerably above the sea-level. The Welland Canal, constructed by the British Government, on the Canada side, passes around the Falls of Niagara, and opens a way for navigation, through which the vessels that trade between Chicago and Liverpool find their way.

Lake Huron is the deepest of all the lakes.

The Lake Superior copper-mines are in Michigan. They bring shipping to Ontonagon and Eagle Harbor. This State is next to Pennsylvania in its manufacture of iron; it is rich in agriculture, and next to New York in the production of salt. There are extensive salt-works in Saginaw Valley.

Michigan has but few rivers or mountains of any consequence.

Much of the State is heavily timbered, and the timber trade is therefore very valuable, as there is so much lake coast from which it can be easily floated to market in any direction. The steamers on the lakes consume immense quantities of wood that is cut in this State.

Moreover, the lake frontage of Michigan, you will be surprised to find, is more than 1000 miles in length, and greater than the sea-front of any other State in the Union, except Florida.

WISCONSIN.

This is the youngest of the five States into which the magnificent land-grant made by Virginia to the United States, in 1787, has been erected. You see by examining the map that the headwaters of the Wisconsin river almost join those of the Fox, one of the lake tributaries. In former days, and when the country was a wilderness, the Indian traders and trappers used to pass this way in their canoes from the lakes to the Mississippi. They had to carry their canoes overland only a short distance. The portage was near the place where now stands the city of Portage.

Wisconsin is level, or rolling; it has no mountains, and there is in it but little land unfit for cultivation.

This State, more than the other four, abounds in lakes. Madison, the Capital, is beautifully situated in the midst of a nest of them.

Prairie-du-Chien (-*sheen*), on the Mississippi, is one of the most beautiful of prairies.

The agricultural resources of Wisconsin are the same as those of Michigan and her sister States. The timber trade of the State is very large.

Wisconsin has also valuable lead-mines.

The west winds of winter sweep across this State from the land, and are cold; on their passage across the lake to the Michigan shore, they are warm. Hence, though in the same latitude, Wisconsin is colder than Michigan.

The Great Lakes.—These border chiefly on the Western States, and separate them from the Dominion of Canada.

You now understand (see the Orographic View of the United States) how the Falls and Rapids between Lake Superior and the sea, show that the Lakes are situated upon three terraces, one above the other, and in such a manner that, in going from the sea to Lake Superior, you ascend by three steps. The first lands you on the Lake Ontario terrace ; the second, above the Falls of Niagara, where lie the three middle Lakes, and the third, above Sault St. Marie, on Lake Superior, at least 600 feet above the sea-level. On a plateau 1,000 feet above this terrace, both the Mississippi and the Red River of the North take their rise. 1.600 feet

STORM ON LAKE SUPERIOR.

therefore is the total descent which the waters from this plateau, and 600 feet the descent which the waters from this terrace, have to make before they get to the sea.

Now, by observing the falls of the rivers, as they come from their sources on their way to the sea, you may trace the shapes of the terraces into which nature has arranged our country. The Welland Canal passes on the Canada side, around the Falls of Niagara.

The Sault St. Marie (*soo sent ma'ry*) between Lakes Huron and Supe-

rior, and the Falls of St. Anthony on the Mississippi, 12 miles above St. Paul, in Minnesota, indicate the limits in these portions of the Lake Superior terrace.

The commerce on the Lakes is very great; it is chiefly domestic, but it greatly exceeds in value the whole foreign trade of the country. With canal-boats, lake-steamers, and sailing-craft, the Great Lakes, during the navigable season of seven months, give employment annually to not less than five thousand vessels.

The storms on the Lakes are as furious as those at sea, and the waves that they raise are as violent. The Storm-warnings, now issued by the Government Signal-Service and Weather-Bureau, from Washington, are greatly needed on these stormy waters; and in several instances already, the timely telegram, announcing the tempest's approach, has been the means of staying the departure of ships and steamers from port, and of saving many lives and property of immense value.

MINNESOTA.

Minnesota has hills, but no mountains. Its soil is a rich and black loam, that is very fine for wheat and grass.

"The Father of Waters" takes his rise from Lake Itasca in this State. It abounds in lakes, many of which are beautiful, though none of them are large. The Falls of St. Anthony and of Minnehaha are noted for their beautiful scenery.

Minnesota lies partly on the northern slope of the great "Divide," and its summit is crowned with lakes of clear water. Its summer climates are delightful, especially to persons with pulmonary diseases ; but its winter climates are very severe ; they are colder than the winters of Wisconsin, because the country is higher, farther north, and more exposed. None but the hardiest plants can withstand them.

Pembina (*pem'bi-na*) is a frontier town where the half-breeds and others from the Red River settlements of the North come every summer to trade their furs, peltries, buffalo-robes, tongues, and pemican. These people also go down to St. Paul annually in large caravans with the spoils they have won from the chase.

The forests of this State abound in fine timber. Vast quantities of it are cut and floated down to the river markets below in huge rafts, on which the rafts-men build their shanties and live. The timber-trade is an important branch of business all along the Upper Mississippi and its tributaries ; much lumber is produced by the mills at St. Anthony, which are driven by the fine water-power there.

The limits of Minnesota extend as far north as the parallel of 49°, and farther than those of any State east of the Mississippi.

MISSOURI AND IOWA.

The climates, soil, and productions of Missouri and

Iowa correspond to the soil, climates, and productions of Kentucky, Ohio, and Illinois, which are between the same parallels but on the opposite side of the Mississippi, and which have already been described.

In Missouri, tobacco cultivation receives attention, as it does in Kentucky, and the vineyards occupy laborers, as they do in Ohio. The wines of Missouri are excellent. This State was a part of the Louisiana purchase, and among her other sources of wealth she boasts of a mountain of iron.

St. Louis was settled by the French. Geographically, this city is the commercial centre of the Mississippi Valley. It is an entrepot of great importance between the East and the West, and has an extensive inland commerce.

Midway between the Atlantic and Pacific Oceans, St. Louis is on the wayside of the great inland routes for trade and travel. The Western States alone have 35,000 miles of railway connection.

Situated near the meeting of the two great rivers in the northern hemisphere, and with a back country embracing thousands of miles of the most fertile lands, and containing in its hills veins and deposits of gold and silver of fabulous richness, she is in the focus of a vast and expanding trade. No city in the world has a more dazzling future than St. Louis.

The Railroad to the Pacific has placed St. Louis in commercial connection with the marts of that ocean, and the early completion of the railway-bridge now in process of construction across the Mississippi, will make the connections of that city, by rail, complete from East to West, as they already are, by water, from North to South.

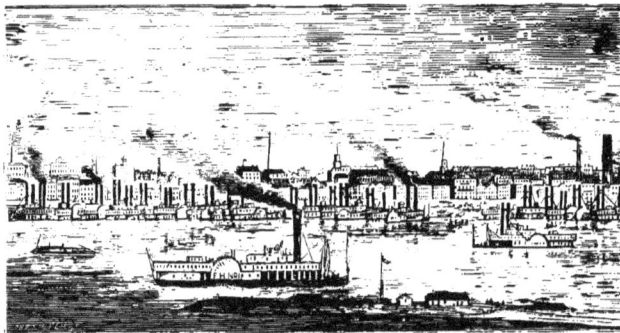

ST. LOUIS.

Iowa is a level State. It has no mountains, but hills enough to divide it into watersheds, and with inclination sufficient to give the country wholesome drainage and abundant water-power. Its lands are fertile, yielding bountiful harvests of wheat and corn.

Dubuque, on the Mississippi, is the centre of a growing business.

In some parts of this and other Western States, coal is often so dear and fire-wood so scarce that the people use corn as fuel. The great expense of sending their grain to market makes it cheap.

So great is the expense of getting the raw produce of the West to the Atlantic seaboard, that the engineer officers of the army who have been sent out there, report that, by present routes, corn grown 100 miles west of Chicago cannot pay the expenses of carrying it to New York.

The study of geography, as it bears upon questions of political economy, is highly instructive and profitable, and facts like these are important. So also are the routes of commerce, because these touch the prosperity of all of the States in an eminent degree.

KANSAS AND NEBRASKA. (Map. p. 73.)

The surface of these States consists of barren wastes, rolling prairies, and grassy plains, with borders and clumps of timber along the streams and in the bottom-lands.

Bears, deer, wolves, and buffalo abound in many parts.

A very small portion of the country has been reduced to cultivation as yet, but the soil and climates are suitable to the great staples of the States in corresponding latitudes east of the Mississippi river. These States are also without mountains.

The best lands in Nebraska are in the eastern half, which is the most thickly settled. The western half is more suitable for pastoral life.

Omaha is a flourishing and growing city.

The Platte, like all the western tributaries of the Missouri, during summer and autumn, when droughts prevail, often presents the singular spectacle of a river in flow near its sources and its mouth, but without any continuous stream in the intervening portion of its bed.

Its head-waters are fed by the snows, and lower down in the valley the rains, the rivers, and lateral tributaries are sufficient to keep up the current; but as it crosses the "mild winter belt" the water is either absorbed by the earth or sucked up by the sun, and at this season in this part of the country many "dry creeks" are found, which at other seasons of the year are dashing rivers.

The soil in the valley of this river is fertile, and is very productive when there is water.

The climate of Kansas is mild, and the winters are not of long duration, nor of great severity.

The staple is Indian corn. Gypsum and coal abound in the State.

The eastern part of the State has been rapidly settled; the western part still contains some Indians.

Leavenworth is the metropolis of the State.

DAKOTA TERRITORY.

This territory, though it lies between the same parallels of latitude with Minnesota, differs greatly from it in climate; it is not as well watered, its rain-fall is not as great, neither is it as abundantly supplied with lakes; it is a much drier country than Minnesota.

The buffalo still swarms over its plains at certain seasons, and hunting and trapping is an important branch of industry among the hardy settlers of this territory, who furnish us with bear-skins, buffalo-robes, hams, and tongues.

COLORADO, WITH WYOMING AND MONTANA TERRITORIES.

This region, one State and two Territories, though lying chiefly on the eastern side of the Rocky Mountains, nevertheless occupies on both sides portions of the great watershed which separates the waters of the Atlantic from those of the Pacific; for you observe that both the Colorado and Columbia rivers have their head-waters, the former in Colorado, the latter in Montana. These two rivers eventually find their way through the water-gaps in the mountains to the Pacific. (See the Orographic View of the United States.)

The crest of this dividing ridge is from 5,000 to 6,000 feet above the sea level, though some of its peaks have more than twice that altitude. These high peaks are always covered with snow, and it is the melting of the snows of these mountains in warm weather which feeds the rivers on the eastern slopes and prevents them from running dry in summer.

The head of navigation of the Missouri—3,000 miles above St. Louis—is at Fort Benton, in Montana. Large numbers of steamers go up there every season, bringing gold from the mines of the settlers.

In the centre of Montana are the great falls of the Missouri river, among the most picturesque in America.

The lovers of the chase frequently come over from Europe, and taking an Indian for their guide, spend the summer in the western prairies, hunting buffalo and other large game.

In the highest parts of this very mountainous region, the snow lies on the ground all the year. The precious metals abound, and the mines of silver and gold are worked with great profit.

There is a peculiar climate along this part of the eastern base of the Rocky Mountains. It may be called THE MILD WINTER BELT; for when the plains of Minnesota, Dakota, Nebraska, Iowa, and Missouri, are covered with snow so as to deprive the wild cattle of subsistence, the buffalo finds abundant pastures in this mild belt, where it passes the greater part of the winter. Thus you see that, although in the same latitude as the States east, and higher above the sea-level, yet here, at the base of the mountains, you have winters so mild that the lizards and reptiles of Texas are found here; and the winters at Fort Laramie, on the head-waters of the Platte, in Wyoming, are so much milder than they are at St. Louis, that the river, at the fort, does not generally freeze until long after the navigation has been closed at that city.

Fort Laramie is not only higher than St. Louis above the sea-level, but it is further north: for both of these reasons the winters at the fort ought to be the colder; but the winter rains make them milder.

It is well for the geographical student to be acquainted with these facts; but he should not be content with that; he should strive to understand their cause.

Winter is the season of the heavy rain-fall on the Pacific slopes of the Rocky Mountains. It may be called THE MILD WINTER BELT; and it is a law of nature that when vapor enough is condensed to make a gallon of snow-water, heat enough is liberated and set free in the surrounding air to boil nearly six gallons of water. It is the heat from this source that tempers the winter climates all along the eastern slopes of the Rocky Mountains and the plains many miles to the eastward.

With this explanation you arrive at a knowledge of this fact touching the climates of the trans-Mississippi country—viz., that as you go west, the winter climate grows milder, till you reach the Rocky Mountains.

Of all the States and Territories, Wyoming is the only one where women are allowed to serve on juries.

The soil about Cheyenne is very rich, and, when irrigation can be had, never fails to produce abundantly. Coal has been found in this neighborhood.

In the same vicinity, also, is the picturesque and unique scenery called the "Garden of the Gods," through which a beautiful stream is constantly flowing, and into which you enter through a natural gateway cut out of the solid rock.

In the territory of Dakota there is a district covered by large masses of indurated clay and marls, which have been worn by the weather into architectural forms and fantastic shapes. In this district are the "Bad Lands" of Dakota.

Colorado is famous for the deep and enormous Cañons (*kan-yons'*) or gorges, which mark the line of the Colorado river.

CAÑON OF THE COLORADO RIVER.

Colorado embraces within its borders a remarkable system of mountain peaks, and the head-springs of four large rivers, namely,—the Colorado, flowing west into the Pacific; the South Platte, flowing in the opposite direction to join the Missouri; the Rio Grande; and the Arkansas.

Questions.—How, geographically, is Michigan State divided?—Why is its climate so mild?—What is the freezing-point of fresh water?—How long is navigation suspended on the lakes annually?—What tempers the cold west winds of winter as they approach the eastern peninsula of Michigan?—Which is the largest of the great lakes?—Why do you suppose it to be higher than the rest above the sea-level?—How do vessels get from it to the lower lakes, and back?—How many steps or levels do the Lakes occupy?—Which lakes are on the middle level?—What canal passes around the Falls of Niagara?—On which side of the river is the Welland Canal?—Suppose you knew the fall of each river or strait between Lake Superior and the sea, could you tell the height of each lake above the sea-level?—Which is the deepest lake?—Where are the Lake Superior copper-mines?—To what lake towns do they give importance?—In what branches of industry does Michigan vie with New York and Pennsylvania?—Where are her salt-works?—Describe the face of the country in this State.—How does Michigan compare, as to the length of her shore-line on the lakes, with the Atlantic States?

What, on the map, are the most striking geographical features of Wisconsin and Michigan?—Between what rivers, which empty their waters into the Mississippi, and others which empty into the lakes, is there a short portage?—Describe the face of the country in Wisconsin.—Which State can boast of the greatest number of lakes?—In what important branches of industry, besides agriculture, are the people of Wisconsin engaged?—Why is the timber trade so important in prairie countries?

Upon how many terraces do the great lakes lie?—How high is the Lake Superior terrace above the sea-level?—How can you find out the edges of this terrace?—How, as to magnitude and importance, does the commerce of the lakes compare with the foreign trade of the country?—What is the total navigable length of the Mississippi and its tributaries?—How many vessels, including canal-boats, are annually engaged with lake commerce?—Are the lakes vexed with storms as the sea is?—Describe the face of the country in Minnesota.—What are its chief agricultural productions?

What States, east of the Mississippi, do Missouri and Iowa resemble? (See page 63.)—What agricultural staples, besides wheat and corn, are cultivated in Missouri?—What famous mountain is in this State?—By whom was St. Louis settled?—What geographical circumstances make the site of this city so important?—Describe the advantages of its situation.—How many miles of railway have the Western States?

Is Iowa a mountainous State?—Describe the face of the country.—Its commercial relations.

What great river rises in Minnesota?—In what lake?—Where is the head of navigation on the Mississippi?—Upon what two great watersheds does the State of Minnesota lie?—Describe its climate.—Can you explain why the winters here are so much colder than they are in Wisconsin?—How far north does Minnesota extend?—What trade is actively carried on in Pembina?—With whom?—To what branch of business do the forests of this State give rise?—What famous falls are at St. Anthony?—Its water-routes to the markets of the Atlantic seaboard?—What are its staple productions?

Describe the face of the country in Kansas and Nebraska.—What wild animals are found there?—With the climates of what States would you compare the climates of these? (Refer to the Orographic View of the United States.)

Between what parallels of latitude does Dakota lie?—How does its climate compare with that of Minnesota?—Describe its natural aspects.—What still constitutes an important branch of industry to the settlers of this Territory?

Do Colorado, Wyoming, and Montana lie wholly within the Valley of the Mississippi?

How high above the sea-level is the crest of the great watershed between the two oceans?—How high the loftiest peaks?—To what point is the Missouri river navigable?—How far is Fort Benton above St. Louis?—How long on the Rocky Mountains does the snow lie?

In what other minerals besides the precious metals does this section abound?—What is a peculiarity of its winter climates?—How far does the mild-winter belt extend?—What proof can you give of the mildness of the winters in this belt?—Can you explain why the climate is milder here than it is several hundred miles to the eastward, and in the same latitude?—At what season of the year does the great snow-fall take place in the Rocky Mountains?—What is said of the soil about Cheyenne?

WESTERN STATES

(For West Virginia, see p. 49 ; for Kansas, Nebraska, and Western Territories, see p. 73.)

Scale of English Miles.

WINNIPEG LAKE

JAMES BAY

B R I T I S H   A M E R I C A

LAKE OF THE WOODS

LAKE SUPERIOR

LAKE MICHIGAN

LAKE HURON

LAKE ERIE

M I N N E S O T A

W I S C O N S I N

I O W A

I L L I N O I S

M I S S O U R I

K A N S A S

I N D I A N   T E R.

A R K A N S A S

T E N N E S S E E

K E N T U C K Y

I N D I A N A

O H I O

M I C H I G A N

W E S T   V I R G I N I A

V I R G I N I A

N O R T H   C A R O L I N A

SPRINGFIELD

NASHVILLE

COLUMBUS

LESSON XXXVI.

Studies on the Maps of the Western States and Territories. (Pp. 67, 73.)

Extent and Boundaries.—Name the Western States and Territories (p. 58).—Between what parallels of latitude do they lie? (Map, pp. 32, 33).—How are they bounded, as a whole, on the north?—On the south?—Between what meridians of longitude do they lie?—How do you bound them on the east?—How, on the west?

What Western States border on Lake Erie?—What, on Lake Huron?—What, on Lake Superior?—What, on British America?—On the Ohio river?—On the Mississippi river?

Bound each of these States and Territories separately.

Judging by the eye and the map-scale, which is the largest of all the Western States?—Which is the smallest?—Which has the most Lake coast?—Which has the greatest extent of border on the great rivers?

Rivers.—Trace the three great rivers of the Western States: the Mississippi; the Missouri; the Ohio.

Across what, and between what States and Territories does the Missouri flow?

What river flows between Indiana and Illinois?—What two rivers from Tennessee traverse Kentucky?—How far above its mouth do they empty into the Ohio?—How near do they approach each other?—Which empties into the Ohio farthest up?—Where is the Muskingum river?—Where does the Cumberland rise?—Describe the Licking, the Kentucky, and Green rivers.—Name the rivers of Ohio.—Describe their course, and tell where they empty.—What rivers empty into the Mississippi from the Western States?—Does the Minnesota river empty above or below St. Paul?—Are most of the rivers of Minnesota drained into the Missouri or Mississippi?—Is New Madrid above or below the mouth of the Ohio?—Does the Missouri empty into the Mississippi above or below the mouth of the Ohio?—Where does the Kansas river empty?—Is this above or below Jefferson City?

Where is the Red river of the North?—Its source?—Its course and terminus?—Where are the head-waters of the Missouri?—Which way do they flow from their source for a considerable distance?—Where does the Des Moines river rise?—Where does it empty?—Trace the Kansas river.—What rivers, which find their way into the Pacific, flow from Colorado, Wyoming, and Montana?

Where does the Platte river rise?—Its course?—Its mouth?—Through what States do the Kansas and Platte rivers run?—Where is Smoky Hill Fork? (See Kansas, p. 78.)

Lakes.—Describe the lake-basin of the Western States.—What rivers connect Lake Huron and Lake Erie?—What lake lies between them?—What is the river that flows into it?—How are Lakes Huron and Michigan connected?—Lakes Huron and Superior?—Erie and Ontario?—Ontario and the sea?—How long, judging by the eye and the map-scale, is Lake Michigan?—How wide?—How broad is Lake Superior?—Do most of the rivers of Minnesota rise in lakes?—Which is farthest to the north, Lake Traverse or Lake Itasca?—Where is Lake Pepin?

Bays and Straits.—Where is Green Bay?—Saginaw Bay?—Keeweenaw Bay?—Georgia Bay?—Where is the Strait of Mackinaw?

Routes and Distances.—What is the course and distance from Wheeling to Baltimore?—From Wheeling to Parkersburg?—To Charleston?—How far is it from Charleston to Point Pleasant at the mouth of the Great Kanawha river? Map, p. 45.

How far is it, going by steamer, from Charleston to Cincinnati?—To Louisville?—To St. Louis?—To New Orleans?—Find Lexington, Ky.—How far is it from Louisville?—From Cincinnati?—How far is Louisville from Nashville?—From Memphis?—Is Louisville above or below Cincinnati, on the Ohio?—How far is it from Louisville to Chicago?—To St. Louis?—How would you go from Chicago to Indianapolis?—From St. Louis to Indianapolis?—How would you go from Cleveland to Dayton?—From Cleveland to Chicago?—To St. Louis?—To St. Paul?—How far from Cincinnati to Lake Erie?—How would you go?—How would you go from Springfield to Terre Haute?

How would you go from Fort Wayne to Terre Haute?—From Fort Wayne to Milwaukee?—How far is Dubuque from Pembina?—How far is Pembina from Omaha City?—From St. Louis?—How would you go from Des Moines to Chicago?—Where is Burlington?—How far above St. Louis?—How far is St. Louis from New York?—From New Orleans?—From Chicago?—From San Francisco?—How could you go from Omaha to Topeka?

Mountains.—What part of West Virginia is most mountainous?—What mountains are there in Kentucky?—Which part of this State is most mountainous?—Describe the mountains of the Western States generally.—Where is Pike's Peak?—Where is Long's Peak?—Where is Iron Mountain?—Are there any mountains in Dakota?

Watersheds.—Describe the watersheds of the Western States and Territories generally.—State, in order, how each one of these States and Territories is drained.

NOTE.—The scholar will do well to examine the Orographic View of the United States, at this point.

Mines.—Where are the lead mines of Missouri?—The iron mines?—The lead-mines of Illinois?

Chief Towns.— Point out Wheeling.—Charleston.—Point Pleasant.—Parkersburg.—Louisville.—Frankfort.—Covington. — Henderson.— Newport. Paducah.—Danville.

Cincinnati. — Columbus. — Dayton. — Where is Cleveland?—Where is Toledo? — Chillicothe? — Marietta? — Steubenville? — Where is Peoria?—Sandusky?—Springfield?—Where is New Albany?—Cairo?—Terre Haute?—Where is Fort Wayne?—Where is Detroit?—Lansing?—How far is Detroit from Chicago?—From Buffalo?—Where is Sault St. Marie?—Where is Dubuque?—Where is Des Moines?—Where is St. Louis?—Is it above or below the mouth of the Missouri?—Where is Jefferson City?—Kansas City?—Topeka?—Lecompton?—Omaha?—Where is Pembina?—Where is Duluth?—What towns in Colorado, Wyoming, and Montana, are at the junction of rivers?—Where is Cheyenne City?—Where is Gallatin?—Virginia City in Montana?—Yankton?—Lincoln?—Leavenworth?—Denver?—Does the Platte river empty above the city of Omaha or below it?

Miscellaneous.—Where are the salt-works of Kanawha river. West Virginia?—Where are the White Sulphur Springs?—What is the only river, rising in West Virginia and emptying into the Atlantic Ocean?—Upon what long vein of rocks are Lexington, Ky., and the Greenbrier country situated?—Where is Fond du Lac?—Isle Royale?—Where are the Pictured Rocks of Lake Superior? *Ans.* In the northern part of Michigan, on the Lake.—Where is Fort Snelling?—How would you go from Chicago to Liverpool all the way by ship?

NOTE.—The Government of the United States has organized a Storm Bureau, for the purpose of rendering the navigation of the Seacoast and Lakes much safer by applying science to the prediction of storms, and establishing a system of signals for warning vessels of their approach and force.

LESSON XXXVII.

The Pacific States and Territories.

TOTAL POPULATION, 835,059.

STATE.	Capitals.	Chief Cities and their Population.
California.................	Sacramento..........	San Francisco .. 149,473
Oregon	Salem	Portland 8,293
Nevada.................	Carson	Virginia City... 7,048
Arizona Territory........	Tucson..............	Tucson 3,224
Utah "	Salt Lake City	Salt Lake City.. 12,854
Idaho "	Boise City..........	Boise City 995
Washington "	Olympia..........	Olympia 1,003
Alaska "	....	Sitka 2,000

1. Coast-Line and Orography.—We come now to the Pacific slopes of our country and the first thing that a geographer does when he reaches a new country is to study its maps and its orography, to learn how the land lies.

The map of the United States tells you that the coast-line of the Pacific, along our southern borders, is not curtained with islands, as is our Atlantic coast in the same latitudes; nor is it indented with deep bays and harbors, as the coast of New England is.

Parallel with the coast there is a range of mountains, as on the Atlantic side. On the Atlantic, this coast-range is separated from the sea by a belt of lowlands, varying in breadth from 50 to 250 miles; while on the Pacific side the hill-country comes down to the sea, and the coast is bluff and steep. Consequently the tide-water country along our Pacific shores is confined to a very narrow belt.

The San Joaquin (*sahn wah-kēn'*) river runs between the Sierra Nevada and the Coast Range.

SAN JOAQUIN RIVER.

2. The Pacific Table-Land.—Between the Nevada and the Rocky Mountains there is an immense table-land or valley, situated several thousand feet above the sea, and varying in breadth from 300 to 700 or 800 miles. This table-land extends all the way from the isthmus of Tehuantepec through British America and Alaska to the Arctic Ocean.

In the widest part of this table-land is the great inland basin of our continent, which is chiefly occupied by parts of Utah and Mexico.

You recognize (see Orographic View of the United States) the various parts of this inland basin by the lakes here and there which have no outlet; such lakes are sometimes salt, sometimes brackish, seldom fresh.

3. Minerals.—The hills and mountains that rise up from this plateau are stored with rich mineral deposits. Silver, gold, copper, and quicksilver, with mines of iron, surpassing in quantity and quality even the celebrated iron mountain of Missouri, have been found in this region. Veins of tin, zinc, lead, and other metals, and beds of salt and soda of unknown extent, are also found.

The chief industry of all this region of country *at present,* is mining; but the agricultural resources are immense, especially in California and Oregon.

4. Climates.—Latitude for latitude, the climates of our country along its Atlantic slopes afford no clue to the climates of the Pacific slopes.

In the former case the winds are from the *land,* and in winter are cold; in the latter case they come from the *sea,* and are warm and moist. This difference of climate *depends simply upon the way the winds blow.* It is so marked, that the seasons in California, instead of being divided into summer and winter, are often alluded to as the rainy and the dry seasons. This is the case all along the Pacific slopes, from California to Chili, except in Peru, where it does not rain at all.

For weeks together in summer not a drop of rain falls in California; her winter is the rainy season; but, as you proceed north, the westerly winds become more dominant, and the rains more copious, so that from Oregon, all the way up to the north, the American slopes of the Pacific are well watered,—whereas from Oregon all the way to Valparaiso in Chili, there is lack of water and a dry season of six months every year.

Oregon and the New England States are in the same latitude. In New England the farmers have to house and feed their cattle all the winter, while in Oregon they lie down in green pastures, and require no shelter.

Here you again perceive that, as a rule, climate mainly depends upon the direction of the prevailing winds.

This rule is one of the keys to Geography, for when you understand the climates of a country, you can judge of its productions, and by its productions you can judge of the occupations and industrial pursuits of its inhabitants—and by their pursuits you may form some idea of their general character.

Questions.—1. What difference do you observe on the map between the coast-line of the Pacific and Atlantic States?—Where are there most harbors, in New England or in California and Oregon?—Which has the most tidewater country, the Atlantic or Pacific States?—Why is the tide-water belt of the Pacific so narrow? *2.* Describe the table-land between the Sierra Nevada and the Rocky Mountains.—How high is it?—Point out on the map some of the inland basins that are situated upon it: How do you tell an inland basin?—Are its lakes fresh or salt? *3.* What are the minerals that are found here?—What is the chief industry of this part of the country? *4.* The country here lies between the same parallels that some of the Atlantic States do: why can you not judge of the climates of the former by those of the latter?—Upon what does the difference depend?—How are the seasons along the Pacific coast generally divided?—Why is not the year in Peru divided into the rainy season and the dry,as well as in Chili and California?—What is the dry season in California and Oregon?—Oregon and the New England States are in the same latitude; contrast their climates.—What, by simply knowing what the climates of a country are, can the geographer tell about it?

LESSON XXXVIII.

Pacific States and Territories—Continued.

CALIFORNIA.

This is the oldest of the Pacific States. The Spanish Jesuits established *missions* or settlements in it at an early day. But it was thinly settled, except by the Indians, until it was purchased of Mexico in 1848, for $20,000,000.

Soon after that gold was discovered, and there was such a rush to the rich mines from all parts of the world as had never before been known. The population of California consists chiefly of Americans; but all nations, even the Chinese, are represented there. The agricultural are quite as great as the mineral resources of this State.

RIDING THROUGH THE TRUNK OF A CALIFORNIA TREE.

The soil there produces with an abundance that astonishes even the farmers of the Southern and Western States, and with an excellence that surprises everybody. All the root-crops and culinary vegetables grow there to an enormous size; and the fruits can be surpassed neither in flavor nor size.

Apples, pears, peaches, grapes, plums, cherries, and melons, with the whole list of small fruits, are of the finest quality. So also are wheat, corn, and the other cereals. Tea, coffee, and sugar, cotton, hemp, and tobacco, find congenial climates and suitable soil, along these sea-tempered slopes. California is also a fine grazing country.

The largest and tallest trees that grow have been found in the forests of this State.

California produces wheat remarkable for its hardness, and for that reason it is called the *"maccaroni wheat;"* it can stand the longest sea-voyages without damage, which the wheat of the Atlantic States cannot do.

Immense quantities of delicious fruits from California, especially pears and grapes, are now sent by the Pacific railway to the Eastern States.

San Francisco is situated on the finest bay along the whole coast; already it is the largest seaport town on the American shore of the Pacific.

A line of steamers has been established thence to China and Japan. Thus a way is being opened by which the farmers along the Pacific slopes of the country will find a ready market for all their surplus breadstuffs; for India, China, and Japan are so densely populated that the cry of the people there is now, and for ages has continually been, for "bread." This line of steamers has, with the aid of the Pacific railway, already opened a tea-trade between China and Chicago.

Yosemite Falls, on the river Merced, afford one of the grandest sights on the globe; they are the loftiest in the world; they are *half a mile high*—sixteen times higher than Niagara.

YOSEMITE FALLS.

The only United States navy-yard on the Pacific, is at San Francisco.

There is in this State a number of flourishing towns, such as Sacramento, Benicia, Stockton, Nevada, and Marysville.

OREGON.

This State has copious rains. Its streams abound with excellent fish, and its forests with the finest of timber. It has a valuable lumber-trade. The winters are mild and the summers do not oppress with heat. Its climates resemble those of Southern France along the Atlantic, more than they do those of the New England States in the same latitude.

It is a fine grain country, but its wheat will not bear long sea-voyages like the wheat of California. It therefore will not bear transportation to the distant markets of the world, for at present the nearest grain-markets of the Old World are not less than ten thousand miles off ; hence the inhabitants have turned their attention to stock-raising and wool-growing.

There is a bar at the mouth of the Columbia river which, with certain winds and times of the tide, becomes dangerous to vessels seeking ingress or egress. The whole State is lacking in good harbors, but its commerce, nevertheless, is on the increase, for it has an important inland trade with Idaho and Montana, which is growing as the population of these Territories increases.

The country between the Cascade Mountains and the ocean is the best part of the State, and is most thickly settled ; but its industries being chiefly agricultural and pastoral, Oregon has no large towns.

Its coast-range, like the Alleghanies, but unlike the Rocky Mountains, is clothed with verdure to the top.

WASHINGTON TERRITORY.

Washington is better watered than any of the Territories in this section : it borders upon the Strait of Juan de Fuca (*hwan' day foo'kah*). which unites with the ocean the magnificent bays and harbors that indent the northwest corner of this Territory. They have water enough for vessels of any size, and room for the navies of the world.

The Columbia river as it passes through this Territory has some beautiful falls, which at a future day will be turned to effective water-power.

Railways have already been built around them, which serve as portages for the river-trade that is already springing up between Montana and this Territory.

Lumber is the chief article of export from Washington. Olympia, the capital, stands at the head of Puget Sound.

East of the Cascade Range, Washington is chiefly a grazing country.

NEVADA.

This is a new State ; gold and silver mines constitute, at present, its chief source of wealth. The mines are profitably worked, and there are signs of great prosperity in the villages about them.

Some of its valleys and sheltered places here and there are very fertile ; but it has many salt plains and barren places, and the settlers have not yet begun to turn their attention from their mines of gold and silver to agriculture, or to any other of the many sources of wealth that undoubtedly abound there. Virginia City is the depot of supplies for the mining interests.

IDAHO, ARIZONA, AND UTAH TERRITORIES.

These three Territories, geographically, are high above the sea-level, very mountainous and rugged. With the exception of the mountain-passes and water-gaps here and there, they are separated from the Western States by the Rocky Mountains, and from those that border on the Pacific by the Sierra Nevada range.

These Territories are mining countries, and the settlers are principally engaged in working the gold and silver mines, or in trapping and hunting.

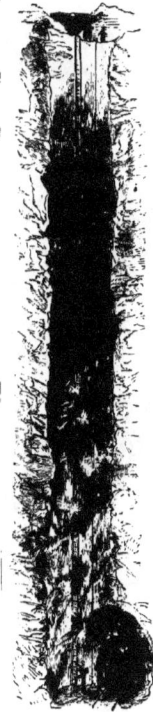

Idaho, as the map shows, is the most northern of the three, and, owing to its distance from the sea, to its elevation and its latitude, its snows are very deep and its winters severe, so that communication in this Territory from one part to another, in winter, is very difficult, and often impossible.

Boise City, the capital of Idaho, is a thriving town.

Arizona, like the other Territories along the slopes of the Rocky Mountains, is rich in minerals, deficient in water, and but thinly inhabited.

It was purchased of Mexico, and is inhabited chiefly by Indians.

SILVER MINE IN NEVADA.

Utah is inhabited chiefly by the Mormons, a sect who style themselves "Latter-day Saints." They profess a new revelation from the Almighty, hold property in common, and practise polygamy.

Salt Lake City is the most important city of Utah.

the same parallels of latitude; which have a sea to the westward, a continent for back country, and about the geography of which observation has fully instructed us. Such a country is found on the Atlantic slopes of Northern Europe, and is comprised within the limits of Western Russia, Denmark, Norway, and Sweden. Here, geographical position and physical aspect correspond with Alaska. This part of Europe, you see, is not so far north or so cold as to be uninhabitable.

The people in these parts of Europe are seafaring. They are largely engaged in commerce and manufacturing, and also in mining. Their forests supply ship-timber and lumber in vast quantities. The soil produces good crops of wheat and barley, turnips, potatoes, mangel-wurzel (a root of the beet kind) and other roots. The time that the cattle have to be housed in winter is of short duration, and the pasturage for the rest of the year is excellent. Stock-raising therefore in these countries, with their long winter nights, is an important branch of industry. Our science therefore teaches that all such industries will, in the process of time, reward the future inhabitants of Alaska.

The Aleutian Islands are volcanic and treeless ; the natives dwell in caves and live upon the bounties of the sea. They depend

SCENE IN SALT LAKE CITY.

ALASKA. (Map, p. 80.)

The acquisition of this Territory has extended the domains of the United States from the Great Manan, an island in the Bay of Fundy, on the eastern shores of America, to the middle of the Arrow Pass, off the eastern shores of Asia, which gives us a breadth of border embracing 126° of longitude. (Map, p. 20.)

It was purchased in 1867. It was called Russian America, and was occupied by fur-traders, who hunted sea-otters, seals, martens, foxes, bears, and wolves. Their chief settlement was at Sitka. The seal fishery is very valuable, and the government of the United States now lets it out by contract.

As yet, not much is known about its geography.

Nevertheless, let us see if already we do not know enough of the principles of physical geography and the laws which regulate climate, to form a tolerably correct idea of this country.

It is in the region of westerly winds.

It has an open sea on the west, and therefore the prevailing winds come from the sea and are warm.

Now let us search the Map of the World for some other shores in the northern hemisphere which, with their adjacent islands, are situated between

for wood from which to make canoes, implements, and utensils of various sorts, upon the drift from Asia, which is cast upon their shores by the Japan current, another mighty river in the ocean, which answers to the Gulf Stream of the Atlantic, and which tempers the climates of these Islands and this Territory, as that stream does the climates of the British Isles, Iceland, and Northwestern Europe.

Mount St. Elias, 17,900 feet high, is the highest mountain in North America. With its everlasting cap of snow, it stands as a landmark between this Territory and the British Possessions.

Questions.—Who first settled California ?—When was it purchased by the United States ?—What caused such a rush of settlers there ?—How, as compared to its mineral, are its agricultural resources ?—Enumerate some of the numerous articles that it is capable of producing.—What kind of wheat grows there ?—What kind of fruit is brought to the Eastern States ?—Describe the city of San Francisco.—Which is the largest seaport town along the entire Pacific coast of America ?—Name some of the most flourishing towns in this State.—For what are its forests remarkable ?

Which is the best watered country, Oregon or California ?—How do the

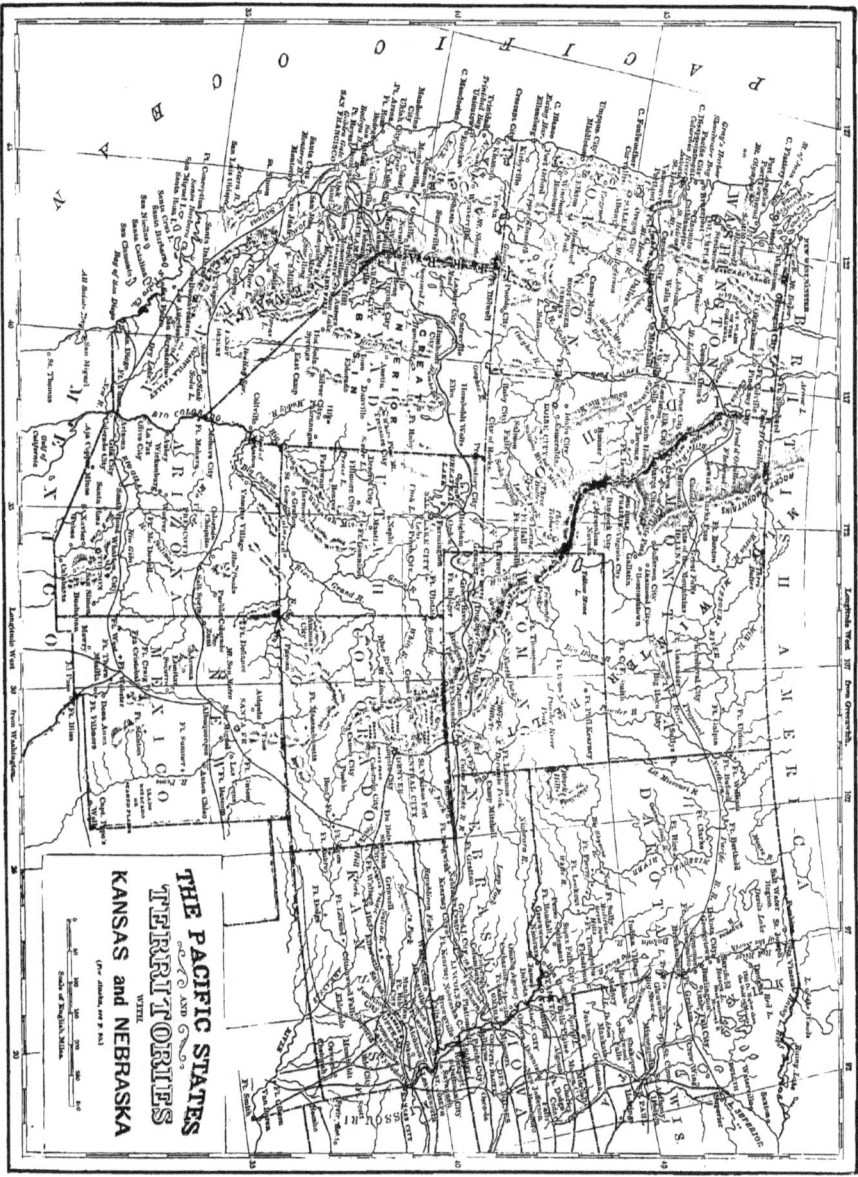

THE PACIFIC STATES AND TERRITORIES

WITH

KANSAS and NEBRASKA

(For Alaska, see P 84.)

Scale of English Miles.

climates of Oregon compare with those of the Atlantic States that lie between the same parallels of latitude?—Does the Oregon wheat stand sea transportation as well as that of California?—What circumstances have tended to encourage grazing in this State?—What part of it is most thickly settled?—How, as to the vegetation growing upon them, do the Alleghanies and coast-range of the Pacific compare with the Rocky Mountains? (Map, p. 73.)

What is the chief source of wealth in Nevada at present?—What is the state of agriculture there?

Describe the geographical position of Arizona, Utah, and Idaho.—Which of them has the severest climates?—What are their chief industries?

Which has the best harbors, Washington Territory or the State of Oregon? —Where are they?

By whom is Utah inhabited?—What is its chief city?

What more can you say of Arizona?—Of whom was it purchased?—Who are its inhabitants?

When was Alaska purchased?—What was it formerly called?—By whom was it occupied, and for what purpose?—Name some of the fur-bearing animals.—Where did the Russians have their chief settlement?—How many degrees of longitude are now occupied by the United States?—Can you name their extreme eastern and western possessions?—What is the area of Alaska? Ans. 577,390 square miles, with a population of 6,000, chiefly Russians.— What is the country good for?—What part of Europe resembles it in position, as well with regard to seas as to latitude?

Describe the Aleutian Islands and their inhabitants.—How do they get wool for their canoes, etc.?—Can you describe the Japan current?—Which is the highest mountain in North America?—Describe it.

LESSON XXXIX.

Studies on the Map of the Pacific States and Territories.

Extent.—How broad and how long is California?—What is the latitude of San Francisco?—What is its longitude?—How much of California is included in the great valley between the Sierra Nevada and the Rocky Mountains?—Bound California on all sides.—Bound Oregon.—Bound Nevada.— Bound Arizona.—Utah.—Idaho.—Washington Territory.—Bound Alaska.— Between what parallels of latitude does Alaska lie?—Between what meridians of longitude? (For Alaska, see p. 80.)

What is the latitude of Salt Lake City?—Its meridian?—What is the latitude of the Aleutian Islands? (See p. 80.)

Rivers.—What two rivers are in California?— Which way do they flow?—What two rivers form the Columbia?—What is the shortest distance between their head-waters?—Measure the length of the Columbia by the scale, and tell its length.

What rivers can you name in Nevada?—Where do they rise?—How do they flow?—Where do they empty?

Describe the rivers of Arizona.—What rivers rise in Utah?—How do they run?—Their source?— Where do they empty?—What rivers rise in Idaho? —Point out the rivers of Washington.—Name the rivers of Alaska.—Point them out.

Routes and Distances.—How far is it from Sacramento to San Francisco?—From San Francisco to Stockton?—To Marysville?

How far is Salem from the Columbia river?—In what direction from San Francisco?—From Virginia City?

What is the length of our Pacific coast?—How far is it from Carson City to San Francisco?—How would you go from one place to the other?—How far from Carson to Salt Lake City?—How far from Great Salt Lake City to Fillmore City?

What is the distance of Tucson from our most important seaport on the Pacific Ocean?—How would you go from Sacramento to Salt Lake City?

Mountains.—Where is Mount Baker?—What and where is the highest mountain in North America?—What can you say of the mountains of Alaska?

Chief Towns.—Name the chief town of California.—Of Oregon.—Of Nevada.—Of Utah.—Of Arizona.—Of Idaho.—Of Washington.—Of Alaska.

Bays, Capes, and Islands.—Name the bays of California.—Its capes.—Capes of Washington Territory.—Capes of Alaska.—Its bays.—Its islands.

Miscellaneous.—Where is the Strait of Juan de Fuca?—Point out Puget Sound.—Where is the Great Salt Lake?—Trace it on the map, and also on the Orographic View of the United States.—Where is the Attou Pass?— How near to Asia and Japan does the border of Alaska bring the United States?—Where are the Aleutian Islands?—What vast, warm current of the ocean sets toward them from the torrid zone? *Ans.* The Japan Stream, called by the Japanese the Kuro Siwo (*ku-ro' se-wo'*) or *Black Stream.*

LESSON XL.

British America. (Map, pp. 80, 81.)

1. Position.—All that you have now learned of the United States and of the general principles of geography, will greatly help you in studying the geography of British America, which, with main-land, islands, and inland waters, embraces an area of 3,500,000 sq. miles.

On its southern border the Dominion shares with the United States the magnificent scenery of the Niagara Falls.

NIAGARA FALLS.

With the exception of Alaska, British America includes all that part of our continent north of the Great Lakes on the one hand, and of the 49th parallel of north latitude on the other. The prevailing winds come from the Pacific side of the continent, which therefore receives the most warmth and moisture.

2. Political Divisions.—The whole of British America is divided into Provinces, viz., Ontario, Quebec, New Brunswick, Prince Edward's Island, Nova Scotia, Newfoundland, Manitoba, and British Columbia.

3. Government.—These provinces or colonies are subject to the British crown, as our "old thirteen" used to be.

Seven of these Colonies, viz., Ontario, Quebec, New Brunswick, Nova Scotia, Manitoba, British Columbia, and Prince Edward's Island, are united in a confederation under the title of "The Dominion of Canada."

Newfoundland, which includes Labrador, has not yet joined the confederation, but, by the Act of Parliament authorizing the confederation, may do so at pleasure. The Northwest and Northeast Territories (Map, pp. 80, 81) are unorganized.

This new Dominion is governed by a Congress and President somewhat as the United States are, but with this difference, that whereas the chief magistrate with us is elected by vote—with them he is appointed by the Crown. His title is *Governor-General of Canada.* Their Congress is called the *Parliament.*

4. Area.—The whole of the British possessions in America to the north of us, embraces an area about equal to that of the United States. Much of it is considered uninhabitable by reason of its severe climate; and so it is *at present.* It is but thinly populated because of the abundance of cheaper and better lands with milder climates in other parts of America.

5. Comparative Geography.—The climate of Siberia is as severe and the winters are as long as in any part of British America; yet in latitudes in Siberia which are considered uninhabitable in America, we find large and flourishing cities and towns.

In British America there is no settlement to the north of latitude 55°; there are only trading-posts.

On the north side of this parallel in Asia (see Mercator's Chart) the cities of Omsk, Tomsk, and Tobolsk, each with a population of 20,000 souls and upward, are found

In Europe we have, very near to the parallel of 60° N. lat., the splendid capital of Russia, St. Petersburg, with more than half a million of people within its walls.

But St. Petersburg, it may be objected, is not a fair comparison, inasmuch as it has a sea to the windward, with the warm waters of the Gulf Stream at no great distance beyond, to supply its prevailing west winds with heat and moisture for the mitigation of the rigors of winter.

Then let us take Moscow. Moscow is too far inland to be affected by the sea; it is in higher latitude than any white man's settlement in British America, and yet has a population of more than 350,000.

Nay, a spot colder than any ever encountered by arctic voyagers is in latitude 64° north in Asiatic Russia. Upon that spot stands and flourishes the Russian city of Yakutsk, with its population of 10,000 souls. There, in winter, mercury freezes, and the temperature sinks down more than 100° below the freezing point of fresh water, and the ground is perpetually frozen to the depth of 60 feet.

Moreover, British America, up to its remotest limits, is already inhabited by Indians, and where the savage can live, surely civilized man, with the vast advantages which knowledge gives him, may also live.

6. The Fertile Belt.—There is in the Dominion a belt of pine country extending from the borders of Lake Superior across to the shores of the Pacific. It is called the "Fertile Belt."

Manitoba and the Saskatchawan river are within the range of this belt. It is in the mild-winter belt already described. (See p. 65.)

Upon it the snow is light, and immense herds of buffalo find pasturage all winter.

It is a good grass, wheat, and barley country.

This belt extends from latitude 49° to 54° 40′ along the Pacific coast, and lies between the same parallels as Labrador.

In Labrador, on the east, you find treeless plains, while with snow for eight months every year, and, except in sheltered spots, you have the limestone rocks, covered simply with a skin of moss, which you can pull off in large flakes, and then the bare surface of the rock is exposed, the soil is so thin.

In British Columbia, on the west, and between the same parallels, the land is covered with soil, and forests of pine and fir which are of a height and girth unknown even in the Atlantic States.

Strawberries and roses grow wild in the fields thereand the ground in winter is seldom covered with snow for more than two weeks at a time.

If you were to travel north from the Great Lakes through the Dominion to the Arctic Ocean, you would remark, at the end of every day's journey, that the trees had become more dwarfed and the vegetation more scanty, until at last mosses and lichens would occupy the landscape.

As the cold weather approaches, the bears and foxes and other animals in those cold countries lay aside their coats of gray, brown, and other summer colors, and put on their winter robes of white. All the animals there are white in winter.

7. The Dominion of Canada contains a population of 3,500,000 whites and half-breeds, and the population of the territories is estimated at 28,700, mostly Indians.

Questions.—**1.** Between what parallels of latitude does the Dominion lie?—From what quarter do the prevailing winds there blow?—On which side, the Pacific or the Atlantic, do they bring most warmth and moisture? **2.** What are the Provinces of British America? **3.** Under what government are the Provinces?—What title includes seven of them?—How is the Dominion governed? **4.** How does British America compare in size with the United States?—Where is it? **5.** Compare the climates of the Dominion with those of Europe and Asia in the same latitude, and tell your conclusions as to the possible future of this part of America.—What animals find pasturage there? **6.** Where is the "Fertile Belt?"—Of what color in winter are the animals in the frozen regions? **7.** What is the population of the Dominion?—Is it most dense on the Atlantic or the Pacific side?

LESSON XLI.

The Provinces of British America.

THE DOMINION OF CANADA.

Population, 3,500,000. Capital, Ottawa, 22,000.

By an Act of Parliament, 1867, the Provinces of Nova Scotia, New Brunswick, and the two Canadas were united under one confederation and called the "Dominion of Canada," and the names of Canada West and Canada East were changed, the former to ONTARIO, the latter to QUEBEC, and permission given for the other provinces to join the confederation. Manitoba, Columbia, and Prince Edward's Island have done so.

QUEBEC.

The Canadas belonged to the French, and were settled by them. They founded the city of Quebec the year after the settlement of Jamestown, Va.

ONTARIO AND QUEBEC.

From the Great Lakes to the sea. the country is dotted with beautiful lakes, which serve in winter. while not frozen over, as so many little furnaces to mitigate the severe cold.

Ontario is a peninsula with a climate like that of Michigan. It has 1,620,000 inhabitants, and is the most populous, productive, and prosperous of the Provinces.

It is a fine wheat country, and its inhabitants during the long winters engage extensively in the lumber business.

The forests abound with the sugar-maple, and the making of sugar from it is another important branch of winter industry.

Coal, iron, rock-oil, copper, and gold are among its minerals. Toronto, with a population of 56,000, is the chief town.

The Province of Quebec lies farther north than the Province of Ontario. It does not get the winter winds, while yet they have in them the warmth they derived in their passage over the lakes, and therefore it has a colder climate than Ontario.

In 1759 a celebrated battle, which gave the Canadas to England, was fought between the French and English before the walls of Quebec, on the Heights of Abraham. General Wolfe and the Marquis of Montcalm, the two opposing generals, both fell, each bravely leading his forces into battle; and in admiration of their knightly bearing and noble gallantry, a single monument now stands on the bloody plain, in honor alike of the vanquished and the victor.

This battle wrested the Canadas from the French and placed them under the British flag, where they still remain. It decided the question which was then trembling in the balance as to which of these two nations should have the ascendency in commerce. That of England is now so great that it employs a fleet of 20,000 vessels.

The majority of the inhabitants of Quebec are the descendants of the early French settlers, who still profess the religion of their fathers, which is Roman Catholic, and use their mother-tongue, which is French.

In consideration of this, the constitution that has just been granted by the Queen of England requires that the Parliament of the Dominion, and the Legislature of Quebec, shall keep their records in both the English and the French languages.

Ontario was settled chiefly by English and Americans. They are Protestants; and English is the language generally spoken there.

Quebec, the capital of the Province of Quebec, with a population of 60,000, is beautifully situated on a high bluff. It is one of the few walled towns on this continent, and is so strongly fortified as to win for itself the name of "The Gibraltar of America."

The scenery around this city is enchanting. The picturesque falls and the natural steps of Montmorenci are near it.

of the moon, the enormous height of sixty and sometimes even of seventy feet above low-water mark. This great rise is effected in the brief space of six hours. The flood rushes in with such force that it has been known to overtake and swallow up, ere they can escape, herds of deer, swine, and other animals that happen to be passing or feeding along the shore. These are the highest tides in the world.

THE FALLS AND NATURAL STEPS OF MONTMORENCI.

Montreal, with a population of 107,000, is the largest city in the Dominion.

Nova Scotia is on the way-side of the great thoroughfare for all vessels passing to and fro between North America and Europe. It is the most eastern point of our continent south of Labrador, and is the nearest to Europe.

Many of the ocean steamers that ply between England and the United States touch at Halifax, both coming and going. The harbor is unsurpassed by any other in America. It is the principal naval station of Great Britain on this side of the Atlantic. English men-of-war are constantly to be seen putting in there for orders, supplies, fresh outfits, and repairs.

Halifax—population 57,000—is, therefore, as you may imagine, a place of much importance. It is a flourishing town, and has a large trade with the United States, chiefly in potatoes, fish, coal, gypsum, and grindstones.

Nova Scotia has a healthful climate, and the timber business is an important one.

Herds of moose and deer are still to be found there.

During the deep snows of winter, these animals fortify themselves against the wolves by ramparts, called "Elk-yards," which they build in the snow.

NEW BRUNSWICK.

New Brunswick, also one of the provinces of the Dominion, has a population of 285,000. Its inhabitants are more maritime in their pursuits than those of either Ontario or Quebec.

The forests of New Brunswick afford abundant supplies of ship-timber, both for the navy and the commercial marine of England.

This province abounds in coal and iron as well as in ship-timber; but the industrial energies of its people are directed chiefly to the lumber business, and the sea fisheries.

St. John, with a population of 30,000, is its chief town.

NOVA SCOTIA.

Nova Scotia, you observe, is a peninsula. The island of Cape Breton belongs to it, and the two together, with a population of 388,000, form the fourth grand division of the New Dominion.

The shores, both of Nova Scotia and New Brunswick, are scoured by the tides of the Bay of Fundy.

These tides rise twice and fall twice, daily; attaining at the full and change

PRINCE EDWARD'S ISLAND.

Prince Edward's Island is of nearly the same size as Delaware. It has 95,000 inhabitants. It is nestled in the Gulf of St. Lawrence, and is protected by Newfoundland from the icebergs that float in the Atlantic. It is sheltered by the highlands of New Brunswick from the west winds of Quebec. Consequently, it has a milder climate than either of these provinces.

The chief industry of the island is fishing and ship-building, tillage and pasturage.

NEWFOUNDLAND

Is cold and sterile; the soil does not yield enough to feed the people, who number 147,000. They derive their means of living mainly from the sea. Fishing and sealing is their chief occupation. One-fifth of the inhabitants of the island reside in the city of St. Johns, which has a population of 25,000.

FISHING ON THE GRAND BANKS OF NEWFOUNDLAND.

The Grand Banks of Newfoundland lie to the eastward of this island. The depth of water upon them varies from ten to one hundred fathoms, and they embrace an area of more than 100,000 square miles. They are one of the treasuries of the ocean, for they are the most extensive and valuable fishing-grounds in the known world.

In the spring and summer of every year, immense shoals of cod, mackerel, and other fish resort there, and are taken in great numbers by the French, English, and Americans.

Seal-fishing on the icebergs, as they drift down along the shores of Newfoundland, commences in the spring.

codfish, split in two, and spread out there to dry; and in that beautiful harbor, ships from all parts of the world, may be seen taking in cargoes of fish.

The celebrated Newfoundland dog attains his most complete development in this island, and is used as a draught animal. Bears, wolves, and reindeer also are common.

It is off the shores of Newfoundland that the cold current which comes down through Davis Straits meets with the warm waters of the Gulf Stream, which flows out through the straits of Florida, and produces those dense fogs which envelop the shores of New England, as well as those of Newfoundland, and which often make navigation dangerous. Many a noble ship has been run on Cape Race because of these fogs, and been lost.

In late summer and in autumn the cold current from the north brings down, in immense quantities, huge icebergs, some of them more than one hundred feet high, and measuring miles in circumference. These are also very dangerous to navigation. They reach hundreds of feet below the surface, and when a ship strikes against one it is like striking against a rock. They often lodge on the Grand Banks, where they remain until broken up by the sea, or melted away by the warmth of the Gulf Stream and the rays of the sun.

One end of the Atlantic telegraph crosses these banks on its way from Valentia Bay, Ireland; it was successfully landed in 1866 at Heart's Content, a small cove of Trinity Bay, Newfoundland.

There are three cables now; one French and two English; the latter have several times been broken.

LABRADOR is under the jurisdiction of Newfoundland, but it is a cold and inhospitable country. Although in the latitude of some of the fairest parts of Great Britain, Labrador has a climate too severe to ripen any of the ordinary cereals; but barley cut when green makes good fodder, and potatoes and a few vegetables sometimes do well.

ST. JOHNS.

In the city of St. Johns, entire acres of ground are to be found covered with sheds that are shingled over with

The country is resorted to in the summer by fishermen and trappers for the sake of its seals and other fur-bearing animals. It is peopled on the north, especially on the Bay of Ungava, by the hardy Esquimaux.

ESQUIMAUX.

MANITOBA AND THE NORTHWEST.

Nearly 200 years ago the king of Great Britain granted a charter to a company of English merchants, called the Hudson Bay Company, which gave it the exclusive right to trap and trade in all that part of British America which lies north of Canada and the United States, and which has lately been annexed to the Dominion.

The beaver, the marten, the muskrat, hare, wolf, fox, reindeer, and bear all abound here, and afford excellent furs and peltries. The Company established various trading-posts throughout these regions, the chief of which is Fort York, at the mouth of Nelson's river, on Hudson Bay.

The new province of Manitoba, which embraces the Red River Settlement, established in 1813 by Lord Selkirk, south of Lake Winnipeg, in what was formerly a part of Rupert's Land, has a population of 12,000 (Census of 1871), made up chiefly of the descendants of the early Scotch and French settlers, and of half-breeds and Indians. It has a Lieut.-Governor, a nominated Legislative Council, and an elective assembly.

The vast Northwest Territory is governed under a separate commission by the Lieut.-Governor of Manitoba.

The country consists of wild prairie, unclaimed forests, and treeless wastes of moss and lichens, rocks, ice, and snow.

BRITISH COLUMBIA AND VANCOUVER ISLAND.

These have been united into one province. They are rich in minerals, pasturage, and timber; Vancouver particularly in coal, and British Columbia particularly in gold, but both in grass and forests. They are in the Dominion.

They have a sea to windward, and though they comprise parallels of latitude that are included within Labrador—which is uninhabitable—their winter climate is so moist and mild that the country is not only inhabitable for man, but the cattle in winter can face the cold without shelter, and can also find green pastures.

These colonies have been recently established. Their entire population, including Indians, does not exceed 50,000; consequently, they have not yet industrial force enough to develop the resources of the country; nor has there been time for industry to adjust itself in regular and permanent channels.

Questions.—What provinces form the Dominion of Canada?—By what Act?—Describe the face of the country in Quebec and Ontario.—Which is the most populous province?—What is its population?—Its productions?—Which is its chief town?—What, its population?—Which is the colder country, Quebec or Ontario?—Who first colonized the Canadas?—What city did they found?—In what year was the great battle there fought, which decided the question of their commercial supremacy?—How large is England's fleet of merchantmen?—What religion do the inhabitants of Quebec still profess?—Why has England required the Assembly of Quebec and the Parliament of the Dominion to publish their acts both in French and English?—By whom was Ontario settled?—What is the prevailing religion there?—Describe the city of Quebec.

What is the population of New Brunswick?—What are the chief branches of industry of its inhabitants?—What minerals has it?—What is its chief town?—What is the shape of the province of Nova Scotia?

What large island belongs to it?—What is the population of the two?—How high do the tides rise on some parts of the shores of Nova Scotia?—Describe the position of Nova Scotia with regard to Europe.—What makes Halifax a place of so much importance?—What is its population?—In what does its trade with the United States chiefly consist?—What wild animals abound on this peninsula?—What is an 'elk-yard?'

How large is Prince Edward's Island?—What is its population?—What is the chief industry of its inhabitants?—What is the population of Newfoundland?—Describe its climate and soil.—How do the people live there?—What are their chief occupations?

Tell where the Grand Banks of Newfoundland are.—Describe them.—When is the fishing season?—What do they "fish" for on the icebergs?—Where do the Newfoundlanders dry their fish?—What is the cause of the dense Newfoundland fogs?—Of what inconvenience are these fogs?—What else besides fogs endangers navigation there?—How large are some of these icebergs?—What is the season for them?—What inconvenience sometimes occurs from the grounding of icebergs on the Grand Banks?—What Province has jurisdiction over Labrador?—Why is it so thinly inhabited?

When was Hudson's Bay Company chartered?—For what purpose?—What animals did they hunt and trap?

What is the new name of Winnipeg?—Of whom does this population consist?—What is their religion?—Describe the general character of the country?—Of what two provinces does British Columbia consist?—What are their natural resources?—They are in the same latitude as Labrador; why are their winters so much milder?—What is their present population?

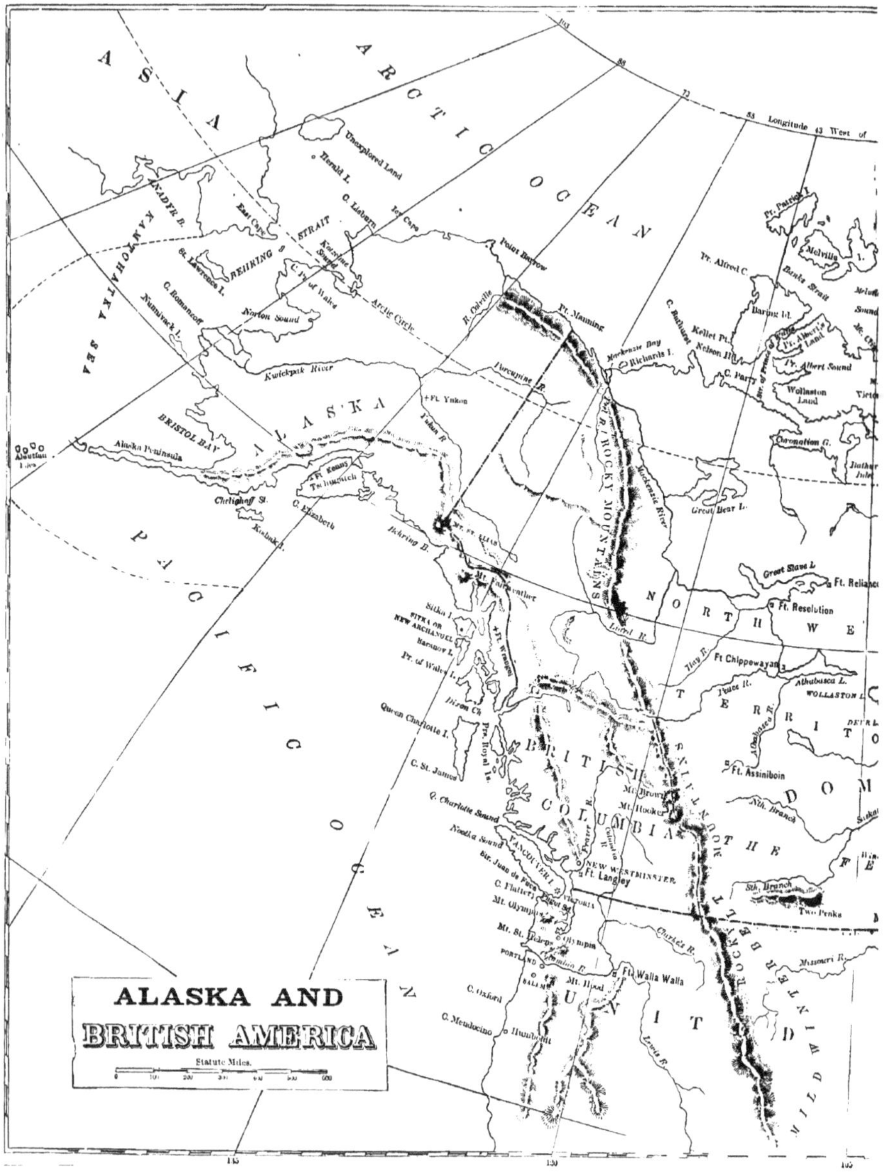

ASIA

ARCTIC OCEAN

KAMTCHATKA

BEHRING STRAIT

Arctic Circle

ALASKA

BRISTOL BAY

PACIFIC

OCEAN

Aleutian Isles

Alaska Peninsula

Point Barrow

ROCKY MOUNTAINS

Mackenzie River

Great Bear L.

Great Slave L.

Ft. Reliance

Ft. Resolution

NORTH WEST

Ft. Chippewayan

Athabasca L.

WOLLASTON L.

Ft. Assiniboin

BRITISH COLUMBIA

TERRITORY

DOM

SITKA or NEW ARCHANGEL

Queen Charlotte I.

C. St. James

Q. Charlotte Sound

VANCOUVER I.

Str. Juan de Fuca

C. Flattery

Mt. Olympus

NEW WESTMINSTER

Ft. Langley

Mt. Brown

Mt. Hooker

WINTER

Two Peaks

Missouri R.

Walla Walla

UNIT

Mt. St. George

PORTLAND

Mt. Hood

C. Oxford

C. Mendocino

Humboldt

ROCKY MOUNTAINS

ALASKA AND BRITISH AMERICA

Statute Miles.

GREENLAND

DANISH POSSESS

OCEAN

Open Polar Sea
Kennedy Channel
C. Constitution
Base of 17 Washington H.
Washington 13 Longitude 7
Ellesmere Land
North Lincoln
North Devon
Jones Sd.
Lancaster Sd.
C. Isabella
Port Foulke
C. York
Melville Bay
Tessiusak
C. Hexburg
C. Bowen
BAFFIN
BAY
C. Adair
C. Hewett
Cockburn Island
Home Bay
C. Walsingham
Cumberland Island
Northumberland Bay
C. Albert
C. Desolation
C. Farewell

C. Paris
C. Brewster
Arctic Circle
ICELAND
Portland

Omenak Fiord
Upernivik
Godhavn
Disco I.
Ritenbenk
Christianshaab
Good Hope
FREDERIKSHAAB
JULIANSHAAB

DAVIS STRAIT

ESQUIMAUX
FOX CHANNEL
Chesterfield Inlet
Southampton Id.
Mansfield Id.
HUDSON STRAIT
Fox Land
Frobisher Bay
Resolution I.
C. Childleigh
Akpatok I.
Ungava Bay
C. Chimo
Nain
Hopedale
C. Harrison
Hamilton Inlet
C. Charles
Belleisle Sr.

LABRADOR

HUDSON
BAY
NORTHEAST
Portland P.
Ltl. Whale R.
Great Whale R.
TERRITORY
JAMES BAY
Ft. Main Ri.
Rupert R.
L. MISTASSINIE
L. MISTISSINIE
R. St. John
Anticosti I.
B. of Notre Dame
NEWFOUNDLAND
Trinity Bay
Grand Banks

OCEAN

C. Churchill
Seal R.
Ft. Churchill
Ft. Severn
C. Henrietta Maria
Nelson R.
Ft. Severn
Severn R.
Ft. Albany
Albany R.
Moose R.
Saguenay R.
NEW BRUNSWICK
Woodstock
FREDERICTON
NOVA SCOTIA
HALIFAX
Breton Louisburg
Sable I.
C. Sable

CANADA
QUEBEC
ONTARIO
Ottawa R.
MONTREAL
OTTAWA
KINGSTON
TORONTO
L. NIPISSING
G. of St. Lawrence
Gaspé
Bay

WINNIPEG
LAKE
Winnipeg R.
Ft. Garry
L. of the Woods
Duluth
L. SUPERIOR
SAULT STE MARIE
Green Bay
L. HURON
Georgian Bay
L. MICHIGAN
CHICAGO
DETROIT
CLEVELAND
L. ERIE
BUFFALO
WHEELING

MANITOBA
Red River of the North
ITASCA L.
Mississippi R.
STATES
L. ST CLAIR
HAMILTON
NIAGARA
NEW YORK
PHILADELPHIA
Hudson R.
WASHINGTON
Connecticut R.
PORTSMOUTH
PORTLAND
AUGUSTA
BOSTON
C. Cod
Long I.

ATLANTIC

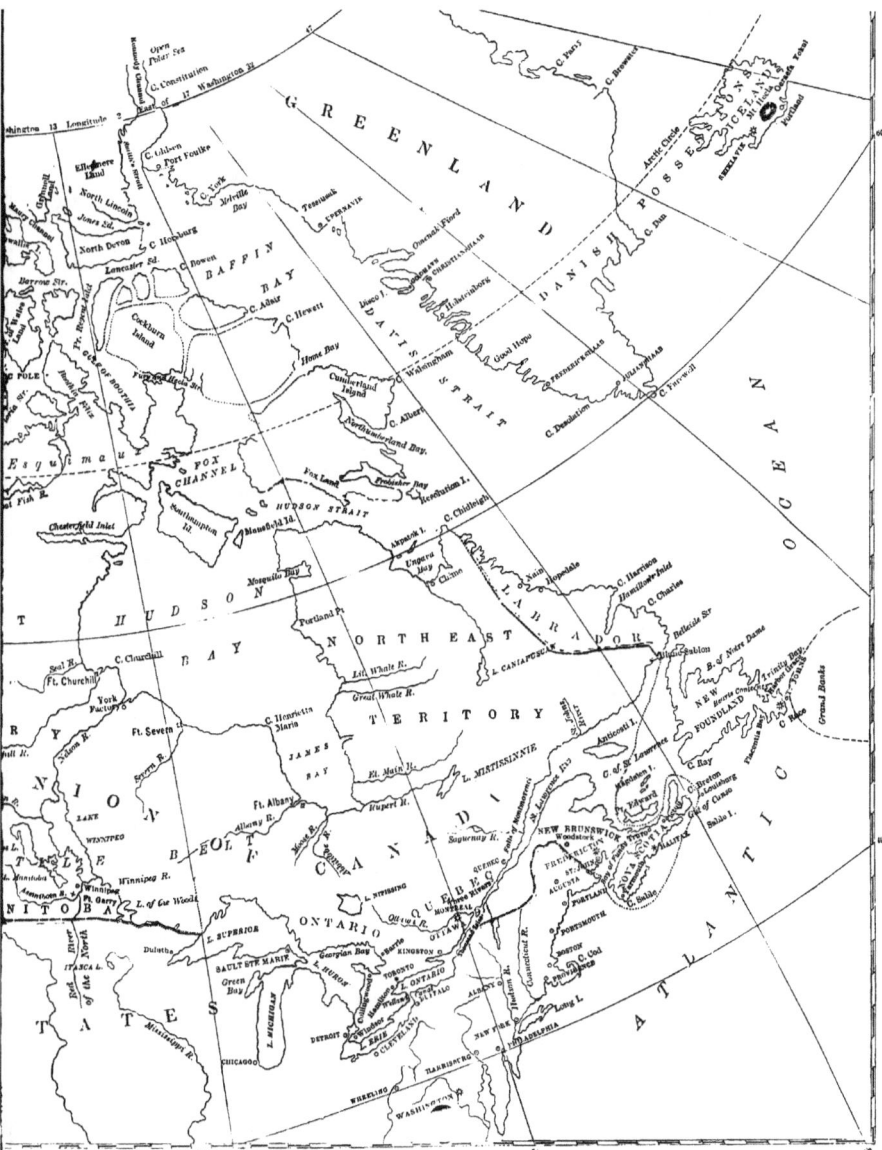

Longitude West 30 from Greenwich

LESSON XLII.

Studies on the Map of British America.

Boundaries and Subdivisions. -Bound British America on all sides. —What great Bay interwashes British America ?—Where are the seven provinces of the Dominion of Canada: Nova Scotia; New Brunswick ; Quebec ; Ontario ; Manitoba ; British Columbia?—Where is Newfoundland ?—Prince Edward's Island ?—Cape Breton ?—Labrador ?—Between what meridians and parallels does British America lie ?—How far is it from Cape Race to Queen Charlotte's Island ?—How far, measuring by the scale, is Melville Sound from the 49th parallel of north latitude ?—What is the latitude of Newfoundland ?

Mountains and Watersheds.—What mountains traverse British America on the West ?—What are their highest peaks ?—How high is Mount St. Elias ?—Where can you find any watersheds in British America ?—*You can always find the watersheds of a country by looking for the sources of the rivers and tracing the outlines of the country drained by them.* (See Wall Map.)

Rivers.—What river separates Quebec from Ontario ?—Where are the sources of the St. Lawrence ?—Its mouth ?—In what direction does it flow ?—What purpose does the St. Lawrence serve? *Ans.* AS A WASTE-GATE STREAM : IT DISCHARGES THE SURPLUS WATERS OF THE GREAT LAKES. (See the *Diagram of the* GREAT LAKES.)—Name all the principal rivers of British America.—Where are the head-waters of the Mackenzie ?—Describe its course and tell its tributaries.—Describe the Red river from source to mouth.—What do you know of the Saskatchawan ?—The Churchill ?—The Nelson ?—The Severn ?—The Great Fish river ?—Where does the Columbia river rise ? What river rises in British America and runs through Alaska ?

Lakes.—What lake lies between Lake Ontario and Georgian Bay ?—Where is Lake Winnepeg ?—Lake of the Woods?—Lake Athabasca ?—Great Slave Lake?—Great Bear Lake ?—Lake Nipissing (*nip'is-sing*)?—From what lake does the St. Lawrence river issue ?—Name the Great Lakes in order.

Gulfs, Bays, and Sounds.—Where is the Gulf of St. Lawrence ?—Where is the Bay of Fundy ?—Where is James Bay ?—Find Coronation Gulf.—Gulf of Boothia.—Where is Baffin Bay ?—Melville Sound ?—Lancaster Sound ?—Fox Channel ?

Straits and Capes.—Belle Isle Strait ?—Davis Strait ?—Hudson Strait ? —Strait of Prince of Wales ?—Cape Sable ?—Cape Race ?—Cape Chidleigh ?—Cape Bathurst ?

Islands. -What islands are in the Gulf of St. Lawrence ?—What islands border it?—Where is Anticosti Island ?—Melville Island ?—Vancouver Island ?—Queen Charlotte ?

Cities.—Where is Quebec ?—Montreal ?—Ottawa ?—Toronto ?—St. John's (Newfoundland)?—Halifax ?—St. John ?—Where is New Westminster ?

Routes of Travel and Trade.—What is the course and distance from Toronto to Ottawa ?—Toronto to Detroit ?—Toronto to Kingston ?—To Montreal ?—To Quebec City ?—What is the route and distance from Quebec to Portland in Maine ?—Which way is St. John from Fredericton ?—Course and distance from St. John to Halifax ?—How far is it from the head of the Bay of Fundy to the Gulf of St. Lawrence ?—How far is it from head of the Hudson river to St. Lawrence river?—How would you go from Halifax to Picton ?—How far from Newfoundland to Labrador ?—How far is Prince Edward's Island from New Brunswick ?—From Nova Scotia ?

Miscellaneous.—What is the area of British America ?—Where is Fort York ?—Where is the *Northwest Passage ?* (Refer to the Trade and Voyage Chart, the last map in the book)—How is Nova Scotia separated from Cape Breton ?—Describe the location of the Grand Banks of Newfoundland.—What is their area ?—Where was the Atlantic cable successfully landed in 1866?—How would you go by ship from Toronto to Cleveland ?

LESSON XLIII.

Danish America.

Danish America consists of Greenland, Iceland, and three small islands in the West Indies.

GREENLAND.

Of the geography of Greenland little is known, except as to the Western and Southern coasts.

It lies just in the midst of the great icy and ice-bearing currents that come out of the Arctic Ocean. It stretches from the parallel of 60° north to an unknown extent toward the Pole. The western shores are rugged, mountainous, indented with numerous fiords and creeks, and fringed with islands.

The lofty interior seems like one immense glacier, and some have supposed that Greenland was made up of a multitude of ice-islands frozen perpetually together.

No trees flourish there. A few culinary vegetables are occasionally raised, but the hardiest cereals have failed.

The population consists of about 1000 Danes, with a native and mixed element numbering 9,000.

The commerce consists mainly in the exchange of the skins of seals, reindeer, and other animals, with eider-down, train-oil, whalebone, and fish, for the comforts of European life.

Godhavn (*god'hovn*), situated on a small island, is the principal place toward the North. Upernavik, in latitude 73°, is, as far as we know, the most northerly permanent abode of man.

Frederickshaab is famous as the place where Otto Fabricius spent the long winter nights translating the Scriptures into the Greenland language.

Greenland, it is supposed by geographers, extends to within a few hundred miles of the Pole. In 1854, Dr. Kane's expedition, after forcing its way over the ice barrier of Smith's Strait, amid intense cold, reached nearly to the high and mild latitude of 83° north.

There they discovered an open and iceless sea, having a temperature of four degrees above the freezing-point of fresh water.

The waves of the sea dashed on the beach with the swell of ocean, and the tides ebbed and flowed. Seals were sporting, and water-fowl were feeding in the open waters.

KANE'S OPEN POLAR SEA.

ICELAND.

Iceland contains an area of 30,000 square miles. Its interior is marked by vast plains of cracked and fractured lava; deep, yawning crevasses; swollen, unbridged streams; deep bogs, and natural steam and vapor baths.

THE GREAT GEYSER.

The ice-mountains called *Yokuls* are volcanoes, occasionally in violent eruption.

The loftiest of these, the Oeræfa Yökul, on the south-east coast, is 6,426 feet above the sea. Mount Hecla, in the interior, is 5,110 feet.

The Geysers (*ghizers*), or Boiling Springs, are, however, the most striking physical feature of Iceland.

The vegetation of Iceland, though dwarfed and scanty, is far better than that of Greenland. Grain will not ripen in Iceland, but garden vegetables are raised, and, along the coast, grass grows in quantities sufficient to sustain the cattle.

Iceland-moss is a valuable article of food, and is exported. Sea-fowl, including the eider-duck, abound; splendid trout are found in the streams, and important fisheries are conducted on the sea-shore. Reindeer run wild in large herds, and the polar bear is occasionally lodged on the island by a cake of drift-ice from the North.

The animal is said to be easily killed, being exhausted for want of food during his voyage from distant shores.

On the eastern coasts, much drift-wood from the Tropics is obtained for fuel; it is washed there by the warm Gulf Stream.

Reikiavik (*rë'kë-a-vik*), the chief town, is a small hamlet.

The population of Iceland is 60,000. The language is Norwegian. The people are fond of literature, and have made from their ancient *sagas* valuable contributions to the history of America, claiming its discovery by them 500 years before Columbus.

Questions.—Of what does Danish America consist?—is much known of Greenland?—Where does it lie?—Its extent?—Bound the three known sides. (See Mercator's Chart of the World.)—Describe its vegetation—Its population—Commerce—Some of its towns

What is the area of Iceland? (Refer to Mercator's Chart of the World.)—Describe its interior.—What are Yökuls?—Describe the Geysers.—Vegetation.—For what is Iceland-moss used?—What is said of game?—Of drift-wood?—Name a town in Iceland.—When do the Icelanders claim to have discovered America?

LESSON XLIV.

Mexico.

1. Mexico, before the Discovery of America.—Before America was discovered by Columbus, Mexico was the seat of a most civilized and powerful empire.

Montezuma was its king; the Aztecs were his subjects.

His splendid Capital, adorned with statues and paintings, stood where the City of Mexico now stands. In it he had groves and fountains, temples, baths, and palaces. His fish-pools, his zoological and botanical gardens were better stocked and filled than any at that time in Europe. Indeed, the idea of a garden of plants, in which the capitals of Europe now boast themselves, was borrowed from Mexico.

The Mexicans of that day had laws, common-schools, institutions of learning, and an academy of science and art. In astronomy they were almost as far advanced as the Europeans.

Their calendar-stone, which has been dug up from the public square in the City of Mexico, showed the movements of the heavenly bodies, and divided the year into months and seasons.

Though not so tall, the base of the Pyramid at Cholula covers four times the space of the grand Egyptian Pyramid of Cheops, which occupies 11 acres of ground.

2. Soil and Productions.—Under that fine climate, the rich soil of Mexico was, when Cortez first went there, in a high state of improvement. The chief agricultural staples were the banana, Indian corn, and the vanilla bean, with the celebrated cocoa, and the lordly maguey or pulque (*pool'kā*) plant, now called the Mexican aloe.

From the cocoa we get chocolate, a beverage borrowed by Europe, and called to this day by the name *chocolatl*, which the Mexicans gave it.

The maguey is peculiar to Mexico. Its leaves served the natives for a natural parchment, upon which the national records were preserved. Its juice, when suffered to ferment, becomes "pulqué," the national beverage of the Mexicans. It is like cider, and when sufficiently "hard" is intoxicating.

The plant is beautiful. Its leaves, six or eight feet long, supply the natives with weather-boarding and shingles for their humble dwellings; its thorns supply them with nails, pins, and needles; its fibre, with strings and cords; and its juice with sugar as well as pulqué. Nor is this all. In some parts its leaves are used in *ice cultivation*. The ice gardens are covered with the leaves, which in the evening are filled with water, like so many little troughs, each holding about a quart. They are porous; also powerful radiators. The radiation from them and the evaporation of the exuded water, bring down the temperature at night to freezing, and in the morning the ice-crop is ready for market.

3. Conquest of Mexico.—In 1519, Cortez invaded the domains of Montezuma, and on his death, he took possession of his kingdom in the name of the King of Spain.

Thus Mexico became a possession of Spain. The Spaniards are still the dominant race there, and their language is the language of the country, and their religion is its religion. They are Roman Catholics, as are all the nations in North America except the United States and the British colonies.

In 1843 Mexico threw off the Spanish yoke, and declared herself independent.

4. Chief Staples.—Sugar, coffee, cotton, hemp, tobacco, cocoa, chochineal, pimento, indigo, wheat, corn, grapes, and olives all find genial climates in this magnificent country, and, when well cultivated, the yield is enormous.

On the plains of Mexico violets are in bloom, strawberries are ripe, and green peas in season all the year.

In the orchards and gardens are gathered the most delicious fruits. The forests abound in ornamental woods and dye-stuffs, and the groves abound with gums and spices, drugs and medicines of much value. Among them may be enumerated the india-rubber tree, the vanilla bean, licorice, sarsaparilla, and jalap—so called from the city of Jalapa (*ha-lah'pah*), where it grows wild.

In Yucatan there is made from a variety of the celebrated Maguey, called henepin, a superior kind of

PULQUE MAN.

as *Sisal* (*sĭ-sal'*) hemp, from
f the Gulf coast, and in the
the south of Tampico and
called Ramie grows finely.
iote (*po-cho'-tä*). It is very
s, and silky fibre.

Caliente.—By looking at the
on will see that in Mexico
ntains on both sides of the
l from the sea by a belt of
dth from a few miles to a

e *tierra caliente*, or hot coun-
nd rife with the pestilence.
the yellow fever is the most
iem.

Cruz, whose citizens, during the sickly
s near but on high and healthy ground.

JALAPA.

Ascending this coast-range of
the height of from 5,000 to
l of the sea, the table-lands
extend entirely across the
o coast-range.

than a broad mountain top, which you
f Mexico; upon which you travel many
rself again in Tierra Caliente, upon the

p, or table-land, is the *Tierra*
te land of Mexico. The cli-

mate is delightful. It is never cold enough to pinch with frost, nor hot enough to oppress with heat.

The city of Mexico is situated upon this table-land, at the height of 7,500 feet above the sea. The houses there are built without chimneys, as the winters are not cold enough to make fires necessary.

The descent from the table-land to the low-land is very precipitous on all sides, but especially on the east, where, if seen from a distance, it appears like a range of mountains. There are only two carriage-roads to it from the Gulf of Mexico, by passes 500 miles asunder: one at Jalapa, near Vera Cruz, and the other at Saltillo, west of Monterey. The table-lands extend to the Arctic Ocean.

7. Mountains.—Its top is not a smooth or level surface, as might be supposed from the word "table-land," but it is diversified with hill and dale, mountain and valley, like other parts of the earth's surface. It has other mountain ranges on the top, some of them shooting up peaks, as Orizaba and Popocatapetl, to the height of 17,800 feet above the sea, and far enough up to reach the regions of perpetual ice. Both of these are slumbering volcanoes, though they are capped with snow.

Orizaba is in sight from the sea, and Popocatapetl from the city of Mexico.

The latter is a solfatara, down into the caverns of which Cortez, during his conquest of the country, sent one of his followers to gather sulphur for the manufacture of gunpowder.

When a volcano ceases to emit flames, and is in the process of extinction, it sends out fumes and gases which deposit sulphur in large quantities; it then becomes a solfatara.

The solfataras of Italy and the Mediterranean supply commerce with most of the sulphur used in the manufacture of gunpowder.

8. The Seasons.—In Mexico, as in all the inter-tropical countries in the world, the seasons are marked by the rains. These commence in June, and last till November.

In Tierra Templada the rainy season is the most delightful, but in Tierra Caliente it is the sickly season.

9. Mines.—This table-land in Mexico, as it is in the United States, is rich in mines of gold and silver, copper, lead, tin, quicksilver, zinc, and iron. Indeed, from Patagonia, all the way up into British Columbia, and in Alaska too, as far as miners have explored, minerals abound, and the richest mines in the world of their kind have been found in this range. It is the metalliferous treasury of the earth.

Silver is the chief article of export from Mexico. Sonora, Chihualua (*che-wah'wah*), and Guerrero (*ger-rä'ro*), are the provinces richest in minerals.

10. Important Cities.—Vera Cruz and Tampico (*tam-pee'ko*) are the chief seaports on the Gulf coast; Guaymas (*gwī'mas*), Mazatlan, San Blas, Acapulco, on the Pacific.

The city of Mexico is encircled by a range of mountains, from which rise two snow-clad peaks—viz., Popocatapetl, 17,800 feet high, and Iztaccihuatl (*ees-tahk-se-hwat'l*), or the *woman in white.* These two giant sentinels stand side by side, lending glory to the landscape, while they impress and charm the beholder

CITY OF MEXICO.

In the climes of perpetual summer, the sight of snow-clad mountains is 'ndescribably grateful. As objects of contemplation they are as pleasing as running water, and as suggestive as the sea.

About two miles from the city of Mexico is Lake Tezcuco, which is connected with the city by a canal, and is the largest and lowest of five lakes in the vicinity. It is salt; the others are fresh.

11. Population.—The population of Mexico, by the official returns of 1869-70, was 9,176,000.

Questions.—*1.* Where, at the time of the discovery of America, was the mightiest empire in the New World?—Who was its ruler?—Where was its capital?—Describe it.—Of what public establishments did the capitals of Europe borrow the idea from Mexico?—What facts can you mention as showing the degree of civilization that existed among the Mexicans?—What was their calendar-stone? *2.* What was the agricultural state of their country?—What their staple productions?—What beverage do we get from the cocoa?—Whence the name?—What plant thrives in no other part of the world except in Mexico?—What is pulqué?—What use is made of the pulqué plant and its various parts? *3.* When did Cortez invade Mexico?—Who was the reigning monarch there?—What became of him?—What is the language of Mexico?—What, the religion?

4. Name the chief staples to which the climate and soil of Mexico are congenial.—What fruits and flowers do you find in season all the year on the plains of Mexico?—Name some of the most valuable drugs and medicinal plants which are indigenous to Mexico.—From what does the medicine called jalap derive its name?—What is henepin?—Where is it grown?—What does it produce?—What is the pochote of Mexico?

5. How is the table-land of Mexico separated from the sea?—Where is Tierra Caliente? *6.* How high is the table-land of Mexico?—How broad is it?—Where is Tierra Templada?—Contrast the climate of Tierra Caliente with that of Tierra Templada.—On which "Tierra" is the City of Mexico?—How many carriage-roads on the east of the table-land?—Where are they?—How far does this table-land extend? *7.* Is the top of it a level country?—What volcanoes have you upon it?—What is a solfatera?—Where are the great sources which supply sulphur for gunpowder? *8.* How are the seasons divided in Mexico?—When does the rainy season commence?—How long does it last?—Which is the sickly season in Tierra Caliente? *9.* Where is the metalliferous treasury of the earth?—What is the chief article of export from Mexico?—Name some of the principal cities in Mexico. *10.* What are the chief seaport towns on the Pacific coast?—Which is the largest city in Mexico?—Name some of the principal towns? *11.* Population of Mexico?

Map Studies. (Refer to Map of Mexico, p. 89.)—Within what parallels of latitude is Mexico included?—Within what meridians?—What are its boundaries on all sides?—What Gulf indents its western coast?—What peninsula between Mexico and the Pacific Ocean?

How far is it from Yucatan to Cuba?—From Yucatan to Matamoras?—From Yucatan to New Orleans?

Where is Tiburon Id.?—Cape San Lucas?—Cape Corrientes?—Cape Roxo?—Cape Catoche?—Where is the Bay of Campeche?—Where is the Gulf of California?—Of Mexico?—Where is the Gulf of Tehuantepec?

Where is the table-land of Mexico?—What mountains run through Mexico?—Name some of the volcanoes of Mexico.—What river forms the northern boundary of Mexico?—Name the other chief rivers of Mexico.—Where is the City of Mexico?—Where is San Luis Potosi?—Monterey?—Presidio del Norte?—Guanaxuato (*guah-nah-wah'to*)?—Merida?—Chihuahua?

How far is it from the City of Mexico to Galveston, Texas?—To Vera Cruz?—To Tehuantepec?—How far from the City of Mexico to San Francisco?—Point out in Mexico the *Tierra Caliente.*—Which are the best mining regions of Mexico?

LESSON XLV.

The States of Central America

The States of Central America all belonged once to Spain. Spanish is the language spoken, and the dominant race is of Spanish blood. Their religion is the Roman Catholic.

These States, with Mexico, occupy the central portion of our continent; they lie between the United States of Columbia, in South America, and the United States of America, in North America.

They derive their importance, not so much from the value of the commerce we have with them, as from their vast natural resources and from their geographical position; for across their borders lie the shortest routes that can be constructed, either by rail or water, between the two oceans.

It was here that Columbus placed the Gates of Ocean which he longed to unbar.

Central America is situated in a belt of volcanic fires that girdles the Pacific Ocean. Izalco, a burning mountain, in San Salvador, was formed in 1770, and has been active ever since. Coseguina is noted for its eruption in 1835, when the air was so darkened by its ashes, even at places 50 miles distant, that friends could not recognize each other, and the fowls went to roost.

These States consist of five Republics and the Balize, a British Province. The Republics, exclusive of Mexico, are Guatemala (*guah-te-mah'lah*), Honduras, San Salvador, Nicaragua, and Costa Rica (*ree'kah*). These are all small States both as to area and population.

The smallest of them in population (Costa Rica) has not as many inhabitants as the city of New Orleans, and the largest of them in area (Nicaragua) is smaller than the single State of Georgia, with only a little more, including Indians and all, than one-third its population.

	Area.		Population.
Guatemala	40,777 square miles		1,180,000
San Salvador	7,334 "	"	600,000
Honduras	47,091 "	"	350,000
Nicaragua	58,167 "	"	400,000
Costa-Rica	21,494 "	"	135,000
Total, 5 Republics	174,863 "	"	2,665,000
Balize (British)	17,008 "	"	25,635
Central America	191,871 "	"	2,693,635

These five Central American Republics, all taken together, are not so large as the State of California, nor as populous, in the aggregate, as the single State of New York.

Their mountains are filled with useful minerals, and richly stored with the precious metals. Their climate, like that of Mexico, is superb; their soil is generous, and their harvest-time lasts the live-long year. Yet these countries are not prosperous.

The soil and climates of Central America are admirably adapted to the production of tea and coffee, cocoa and sugar, cochineal and indigo, cotton and corn, hemp and flax, tobacco and vanilla. Cochineal is an insect.

The forests, like those of Tierra Caliente, in Mexico, abound in ornamental woods, dye-stuffs, gums, spices, drugs, and medicines.

Mahogany of fine quality comes from the forests of Central America.

Cattle of all sorts thrive well.

Numerous mines of silver and gold in the hill-country, lie there ready to be wrought with profit whenever proper energy and skill are brought to the work of development.

These States export a few hides and a little cochineal; some coffee and cocoa, but no tea or sugar, cotton, rice, or hemp.

The largest city in them all is New Guatemala, with an estimated population of 40,000 souls.

The geographical position of Central America is both instructive and important.

Turn to Mercator's Map of the World and study this country.

You observe it connects North and South America, and separates, by a narrow strip of land, the waters of the Pacific from the waters of the Atlantic Ocean.

A ship-canal across the isthmus would do away the necessity of vessels engaged in the coasting-trade of the United States between the two oceans, and save them more than 10,000 miles in the distance to be sailed.

Observe on the map the lakes and rivers of Nicaragua; they seem to offer a favorable route for a ship-canal. (p. 89.)

A railway has been established across the Isthmus of Panama, and it is used as a thoroughfare for passengers and emigrants going to California and China.

Questions.—From what do the Central American States derive their importance?—Name them.—To whom does the Balize belong?—Which of these Republics is the smallest in population?—Which the largest?—How large is it?—Mention some of their natural resources.—Are these countries prosperous?—Name some of the productions for which these countries are adapted.—What do their forests yield?—What do the people there export?—What is the population of the largest city?—In which of the Republics is it?—Describe the importance of their geographical position.—How much would a ship-canal across here, for vessels in the coasting-trade between the Atlantic and Pacific ports of the United States, save?—Where does the most favorable route for a ship-canal appear to be?—What makes it so?—What improvement has been completed?

LESSON XLVI.

The West Indies.

These islands are like stepping-stones across the ocean from Florida to the Orinoco; they are in sight from one to another, almost all the way.

They embay the shores of Central America, and form the dividing line between the Caribbean Sea and Gulf of Mexico on one hand, and the Atlantic Ocean on the other. They keep out the tidal-wave, and make both that gulf and sea all but tideless.

None of these islands, except Hayti, are independent. Those near the coasts belong to the neighboring Republics; but all the West Indies proper, that is, all the islands in the group that lie between Florida and the mouth of the Orinoco, or south of the Bahamas, belong to some European power, and are ruled by governors sent out for the purpose.

Cuba is the largest of them all, and belongs, with Porto Rico, to Spain. They are governed by the Captain-General of Cuba. Martinique and Guadaloupe belong to France; St. Thomas, the island of San Juan, and Santa Cruz to Denmark; Curacoa (cü-ra-so) and St. Eustatia to the Dutch, St. Bartholomew to Sweden, and the whole of the Bahamas, with Jamaica and the greater portion of the lesser Antilles, to Great Britain.

The West India Islands, with the exception of the Bahamas, are all intertropical. They will produce almost anything that the inhabitants choose to cultivate, but their chief staples for export are coffee, sugar, and tobacco, with summer fruits and garden vegetables for the markets of our northern cities.

They have other industries and other sources of wealth besides those which spring from the soil. The sea and the mines, and their cigar factories, are very profitable.

CUBA.

Cuba is called the Queen of the Antilles. It is 720 miles long and averages 60 miles in width. In extent it embraces half the area included within all the islands; it is in size equal to Tennessee.

Havana, with a population of 150,000, is the chief seaport.

A STREET SCENE IN HAVANA

It is the largest and most wealthy city in the West Indies, and has one of the finest harbors in the world. The entrance to it is narrow, and is guarded by the celebrated Moro Castle.

The cathedral in Havana contains the remains of Christopher Columbus.

HAYTI

Once belonged to France and Spain. The island is now inhabited chiefly by negroes and mulattoes, and is divided into the Republics of Hayti and San Domingo.

The town of San Domingo is the oldest of the cities founded by Europeans in the New World. It was established in 1504, and contains 15,000 inhabitants.

This island, like Central America, is unsurpassed in its agricultural resources, and is rich in minerals.

The exports of Hayti consist of mahogany and other woods in the rough, a little cotton, and some coffee.

JAMAICA

Is the third island in size, and belongs to England. Its exports are sugar, molasses, rum, coffee, tobacco, cocoa, allspice, and indigo.

THE BAHAMA ISLANDS belong to Great Britain. Nassau, the capital, is the chief town.

The coral rocks and reefs which skirt these islands on the west are dangerous to navigation. Sponges, sea-shells, and corals are also collected in considerable quantities, and sent to New York and other places for sale.

TURK'S ISLAND is noted for the manufacture of salt by solar evaporation.

ST. THOMAS derives its importance from its fine har-

bor. It is a free port, where vessels pay no duties.

In consequence of this the West India mail-steamers of England and France have made it their place of rendezvous, where they meet the smaller steamers and exchange cargoes and passengers.

BARBADOES is the centre of another English colonial government.

The colored men of Barbadoes (bar-bā'diz) make excellent sailors.

TRINIDAD almost joins South America. It is the largest island in this part of the group, and produces sugar, rum, coffee, cocoa, and ginger.

It is celebrated for a lake of pitch, from which immense quantities are annually taken and carried abroad, and yet there is no perceptible diminution of it. It rises up from the earth as fast as it is taken away. The streets and sidewalks of Paris are paved with it.

The inhabitants of the West Indies are generally of a dark color. They are either brown, like the Moor; red, like the Indian; black, like the Negro; or yellow, like the mixed breeds.

Spanish is the language most generally spoken in all parts of America south of the United States.

The area of the whole group taken together is about twice that of Mississippi, and its population more than four times as great.

There are various clusters of small islands. Chief among these are the Bahamas, pop. 10,000. Of the smaller islands, the most populous are Barbadoes 153,000, Trinidad 85,000, Grenada 37,000. The islands owned by—

		square miles		Population.	
Spain, embraces,		19,177,	"	1,980,000
England	"	15,050,	"	955,000
Hayti	"	28,036,	"	740,000
France	"	1,015,	"	275,000
Holland	"	367,	"	32,000
Denmark	"	120,	"	38,000
Sweden	"	45,	"	3,000

NOTE.—The population of San Domingo, a portion of Hayti, is 136,500.

Questions.—Between what places, on the shores of North and South America, do the West India Islands lie?—Between what sheets of water do they form the dividing line?—Which of these islands are independent?—To whom did they belong?—Which is the largest of them?—Point out those that belong to Spain—To France—To Denmark—To Holland—To Sweden—To England.—Compare Cuba with Tennessee.—Tell its population.—Which is the largest city in the West Indies?—For what is it noted?—Name some of

MEXICO, CENTRAL AMERICA, AND WEST INDIES.

the natural resources of Cuba.—In what zone do the West Indies lie?

By whom is Hayti inhabited chiefly?—How is it divided?—To whom does Porto Rico belong?—What is the capital of the Bahama Islands?—Which are the most populous of the Lesser Antilles?

LESSON XLVII.

Studies on the Map of Mexico, Central America, and the West Indies.

Boundaries.—Between what latitudes are the States of Central America?—Between what meridians of longitude?—Bound these States.—Where is Balize?—Which of these States border only on the Caribbean Sea?—Which on the Pacific?—Between what parallels of latitude do the West Indies lie?—

Describe their situation.—Where are the Bahamas?—How are they situated with regard to Cuba?—Which are the Greater Antilles? *Ans. Cuba, Hayti, Jamaica, and Porto Rico.*—Where are the Lesser Antilles?—Where are the Windward Islands?—Name the two largest of the Great Antilles.

Find Barbadoes.—St. Vincent.—How is Barbadoes situated with regard to St. Vincent?—Find Santa Cruz, San Juan, and St. Thomas. *These three belong to Denmark.*—Point out the Great Bahama.

Bays, Lakes, and Capes.—What Bays and Capes are on the Atlantic side of Central America? What, on the Pacific side?

What lakes can you name in Central America?—In what part is Lake Nicaragua?—What is its size?—How does it connect with the Atlantic Ocean? *Ans. By the San Juan River.*

Name the capes of Cuba.—Of Hayti and San Domingo.—Name any others in the West Indies.—Point out Point Galina.—Cape Hayti.

Mountains.—What are the Mountains of Central America and

the Andes.—How high are they in the Isthmus of Darien? *Ans. Only 750 feet high.*—Has Central America many volcanoes? *Ans. Yes; more in proportion to its area than any other part of the world.*

Rivers.—Describe the Rio Grande; Conchas; Culiacan; Grande de Santiago; Balsas; Coatzacoalcos.

Cities.—Where is the City of Guatemala?—Leon?—Tampico?—Truxillo?—Comayagua?—San Salvador?—Managua?—San Jose?—Kingston?—Brownfields?—Aspinwall?—Panama?—Sisal?

Where is Havana?—Matanzas?—Santiago de Cuba?—Port Royal?—S. Domingo?—Ponce?—Where is the Island of St. Thomas?—Nassau?—Great Cayman?

Routes of Voyages.—How would you go from Havana to St. Thomas?—From Port au Prince to New York?—From Havana to Aspinwall?—From Aspinwall to Vera Cruz?—From Barbadoes to Balize?

What is the distance from Aspinwall to Port au Prince?—How far from Port au Prince, through Lake Nicaragua, to the Pacific Ocean?

ANIMAL AND VEGETABLE LIFE OF SOUTH AMERICA.

LESSON XLVIII.

South America　(Map, p. 102.)

1. Shape and Extent.—South America is triangular in shape and lies partly in both hemispheres, but by far the largest part of it is included within the tropics and in the southern hemisphere.

The *narrowest* part of North America and the *broadest* part of South America lie between the tropics.

As you recede from the Tropic of Cancer toward the North Pole, our continent gets broader and broader ; as you recede from the southern tropic toward the South Pole, South America gets narrower and narrower.

The area of South America is 6,961,864 square miles, and the area of North America is 8,851,728 square miles.

2. Comparative Geography.—The great river of North America runs from North to South ; the great river of South America runs from west to east.

The Mississippi is extra-tropical. With every degree of latitude in its course from north to south, it changes its climate ; and with climate, production and industry vary.

The Amazon is inter-tropical. It runs from west to east. It marks no change of climate, and the variety of production and industry along its banks is such only as is due to the change of height above the sea, as it flows eastwardly from the mountains to the ocean.

Consequently, these two rivers already present conditions for striking contrasts in their geographical relations and commercial aspects.

The lessons of our science teach us to expect that when these two, the most magnificent river-basins in the world, shall be occupied, each according to its capacity, the Mississippi will excel in *way-business*—the Amazon in *through-commerce.*

3. Mountains and Rivers.—The Andes skirt the shores of the Pacific all the way from Patagonia to Panama, and give to the Atlantic slopes of South America a breadth of area that comprises 15-16ths of the whole country.

Consequently, all the great rivers of South America are drained toward the east and empty into the Atlantic Ocean. The waters that are drained off into the Pacific are only mountain-streams that are fed by the melting snows on the western slopes of the Andes. (See p. 96.)

4. Early Civilization.—What Montezuma and the Aztecs were to Mexico, the Indians, under Atahualpa, were to Peru.

What Cortez did to the former, Pizarro did to the latter. He claimed the whole of South America, as Cortez did of North America, for their master, the King of Spain.

The Peruvians had beasts of burden, as the llama ; the Aztecs had none, nor any other domestic animal. The Peruvians had public highways and paved roads. The remains of the great road from Quito to Cuzco, and thence along the plateau of the Andes to Chili, like the Appian Way to Rome, are still to be seen. This road was constructed, for nearly 1000 miles, over pathless heights buried in snow ; galleries were cut for leagues through the living rock ;

ravines were crossed by suspension bridges; precipices were scaled by means of stairways hewn in their steep sides; stone pillars were set up as mile-stones by the wayside to mark the distance. The breadth of this magnificent road was 20 feet; it was paved with heavy flags, and covered in some parts with bituminous cement.

The Peruvians had also temples, fortresses, terraced gardens, and aqueducts ; superb palaces and splendid cities.

In Cuzco was their great temple of the sun, the most magnificent structure in the New World, and in its day far surpassing, for the costliness of decoration, any edifice in Europe. For the royal baths the water was conducted into basins of gold through subterranean channels of silver.

Cuzco, like the city of Mexico, on a table-land, is overlooked by snow-clad mountains, and stands 11,800 feet above the level of the sea.

5. The Three Great Inland Basins—It is worthy of the geographer's notice that there are but three great inland basins in the New World: that of Cuzco, with Lake Titicaca ; that of Mexico, with Lake Tezcuco ; and the great inland basin of North America, which includes Utah, New Mexico, and Nevada, with the great Salt Lake. Of all parts of the continent, these inland basins most abound in the ruins of empires and in the memorials of ancient civilizations.

6. Physical Geography of Intertropical America. —The physical geography of all of intertropical America is alike.

Within this region soil and climate are found that are adapted to all the great agricultural staples of the world. Cotton, sugar, tobacco, cocoa, coffee, tea ; the poppy, the banana, and potato ; hemp of several kinds, flax, cochineal, indigo, wheat, rice, corn, incense, gums, spices, perfumes, drugs, medicines, dye-stuffs, and ornamental woods; with boundless pastures for herds and cattle of all sorts ; all these abound in this favored country as they do nowhere else. Cochineal is a dye ; the insect feeds on cactus plants.

Therefore, in telling of the agricultural productions of one of these intertropical States, we describe the agricultural resources of them all.

Questions.—*1.* What is the area of the two Americas?—Of their Islands? —Compare North and South America.—Show their points of resemblance and contrast. *2.* Compare the largest rivers.—Describe the influence which the course of a river has upon its commerce.—Why should you expect the river traffic on the Mississippi always to surpass that of the Amazon? *3.* On which side of the Andes are its largest rivers?—Into what ocean do they flow ? *4.* How did the Peruvians compare in civilization with the Aztecs of Mexico?—Had the aborigines of America any beasts of burden ? —Who was the ruling Inca when Pizarro invaded Peru ? *5.* Describe the three American Inland Basins, and tell for what they are remarkable. *6.* What is said of the physical geography of intertropical America?—What are the productions of the soil ?

LESSON XLIX.

Equatorial South America. (Map, p. 102.)

THE UNITED STATES OF COLOMBIA.

These States have a population of 2,900,000, and an area of 357,000 square miles. They are, in size, equal to Texas and New Mexico combined.

The United States of Colombia, like British America and the United States, extend from sea to sea, and embrace Panama, with its railway, which is one of the chief sources of revenue to the State, and which, by treaty with the United States, is bound to be neutral in war.

The farmers there grow corn, sugar, coffee, cocoa, and tobacco ; but they produce not much more of these than is required for the scanty home consumption.

Cartagena (*car-ta-jee'na*) was once a great commercial mart of Spain.

A few miles back from it, in the interior, is a group of those curious phenomena, the air volcanoes. They are truncated cones, about 25 feet high, filled with water. The plain on which they stand is about 10,000 feet above the sea.

This city is the principal seaport of the Colombian States. In the colonial times it was a place of considerable importance, but now its tenantless houses and desolate streets show that not much business is done on its wharves.

From Panama we get straw hats and grass hammocks, and these, with a little chocolate, are the chief manufactures that come from these States.

There are no manufacturing establishments of consequence in any part of intertropical South America, except in the fastnesses of the mountains, where transportation of merchandise from the sea becomes very costly.

Bogota, with a population of 45,000, the capital of the United States of Colombia, is situated 8,700 feet above the sea-level. It has two rainy seasons annually, so its climate all the year round is as charming and delightful as the month of May is with us.

Persons at Bogota have only about three-fourths as much atmosphere above them as we have. The diminished pressure caused by this is often, at first, very distressing to strangers. The feeling is like that caused by a shortness of breath. Indeed, owing to the diminished pressure, at great elevations, the blood often gushes from the eyes, mouth, and ears of travellers who ascend the neighboring snow-capped peaks.

In these mountains are found the two natural bridges of the Icouonzo, which span a foaming torrent—one at the distance of 250, and the other 300 feet above.

Silver, gold, and precious stones are found among the mountains in these States.

VENEZUELA.

Venezuela is about three times the size of the Islands of Great Britain and Ireland put together. It contains 2 200,000 inhabitants, including 50,000 Indians, and has an area of 368,220 square miles.

The early Spanish explorers, observing that the natives had built their houses on piles along the shores, called the country Venezuela or "Little Venice."

It is traversed by the Orinoco river. It does not extend as far back as the crest of the Andes, and is therefore less mountainous than the United States of Colombia.

Venezuela is a prairie country. More than two-thirds of it, it is computed, consists of llanos (prairies), upon which immense herds of cattle constantly feed.

These llanos (*lyah'nos*) lie mostly on the left bank of the Orinoco; forests occupy the right bank, and extend thence toward the south, where they mingle with the selvas of the Amazon.

A vast extent of these plains is annually overflowed. In the rainy season the low flat country of the Orinoco becomes, like the borders of the lower Nile, a boundless sea. These waters teem with creeping things: with alligators, reptiles, and the curious fish known as the electrical eel. Horses, when fording the pools, are sometimes knocked down by the latter.

In the dry season verdant plains become barren wastes. The cattle wander for pasture off to the hills; the ponds dry up, and the alligators bury themselves in the mud to hibernate till the coming of the next rainy season, when they may be seen coming up like hideous spectres from the bowels of the earth.

The tides ascend the Orinoco to the distance of more than 200 miles; above that, for a considerable distance, it is navigable.

It was somewhere on the banks of the Orinoco that the vivid imagination of the early discoverers placed the gilded king, El Dorado, and his golden city of Manoa.

The flora of Venezuela is wonderfully rich and varied. A species of mimosa grows wild there, which spreads out its umbrella-shaped top until it attains the enormous proportions of several hundred feet in circumference.

The "cow-tree" is also found in Venezuela; the natives tap it and draw from it a milk-like beverage.

Growing wild in its forest and upon its llanos, there are 45 kinds of medicinal plants, 36 that yield gums and resins, and 240 kinds of trees that yield excellent dye-stuffs or afford fine timber. In addition to these, 180 kinds of useful plants and vegetables are cultivated for food and domestic use.

The inhabitants produce for export, but in quantities by no means large, sugar, coffee, cocoa, tobacco, cotton, and indigo; and their herds yield for the West India markets jerked beef and hides.

Caracas, the capital, with a population of about 40,000, is a city famed for its elegant hospitality.

It stands on the seashore, 3,000 feet above the water, and immediately in the rear is the "*Silla*" (*the saddle*), which, with its two peaks, reaches the height of 8,600 feet. They may be seen many miles out at sea, and are well known landmarks to the navigator.

Caracas was the birthplace of Bolivar, sometimes called the Liberator, the Washington of South America. Varinas is noted for its tobacco.

All this part of the country, as indeed are many other parts of Spanish America, is subject to earthquakes. Caracas was visited and well-nigh destroyed by one in 1812.

Maracaybo is a fine old town, with 20,000 inhabitants. It lies within the air-volcano region spoken of in a former paragraph.

Asphaltum and petroleum in large quantities have been cast up, and the "Lantern of Maracaybo" is a volcano lit up with petroleum prepared in the laboratories of nature. It serves as a lighthouse.

Cumana was the most ancient city in South America.

It declined with the industry of the country, and in 1853 its ruin was completed by an earthquake.

Ciudad Bolivar is situated on the Orinoco, 240 miles above its mouth, and is an active place of business.

It was there that this standard-bearer in the Spanish Revolution assembled the first Venezuelan Congress.

THE THREE GUIANAS.

These three provinces belong respectively to the English, French, and Dutch. They are the only portions of South America that remain in the possession of any European power.

Together they are about the size of California, with an aggregate population of 246,000; viz., British Guyana, 163,000; French, 58,000; Dutch, 25,000. Here the rainfall is greater than it is in any other part of the world, except on the Khasia Hills and at Cherrapungee in India. The Great Kaieteur Waterfall is on a tributary of the Essequibo in British Guyana. It was discovered in 1870. It makes a clear leap of 822 feet where the river is 123 yards broad and 15 feet deep. The surrounding scenery is unique and picturesque, made so by a succession of long, flat-topped mountains rising abruptly from the plain, with precipitous sides like walls of masonry. Of these, Mt. Roraima is the most remarkable. It is 18 miles long and 7,500 feet high. The Essequibo, Orinoco, and Amazon, all have tributaries which take their rise on these singular elevations. They gather strength as they go, and, dashing down the mountain-sides, form a succession of the most beautiful cascades and waterfalls, some of which accomplish a leap of 1,500 feet at a bound.

The coast country is low and flat. It is a continuation of the swamp-belt which skirts the seaboard all the way from the Dismal Swamp in Virginia through Mexico and Central and South America, till you pass the Delta of the Amazon. In all parts of this belt, but especially here, the air is filled with insects, vegetation is rank, and the forests teem with wild dogs, tiger-cats, armadilloes, deer, sloth, ant-eaters, wild boars, raccoons, opossums, etc.; the tree-tops are lively with songsters, gay with parasites, air-plants, and flowers, and noisy with howling monkeys, preaching monkeys, weeping monkeys, and monkeys of various other species.

In these lowlands of intertropical America, as in the jungles of India, the forests are so thick set with trees and undergrowth, and so interwoven with vines, parasites, and air-plants, that the only way of getting through them is to go by the watercourses in a canoe, and, unless you use the axe, you may travel a whole day without finding room on the banks to land, or a place among the trees large enough to light a fire upon. The interior is unexplored.

It was on the upper waters of the Berbice (*ber-bés*) River that the magnificent water-lily (Victoria Regia) was first discovered in 1837. Its leaves lie on the water like broad and shallow dishes; they are large enough to float a child; the flower is fragrant, white, and beautiful.

The Guyanas are inhabited by Negroes, Indians, and Europeans.

The streams and rivers in the tide-water region of this country are deep and sluggish—a sure sign that the country lies low, is flat, and has but little fall for drainage.

The products are sugar, cocoa, and coffee, most of which is consumed in Europe.

The capitals are the largest towns; they are—

Georgetown	Population 25,000
Paramaribo	" 20,000
Cayenne (from which we get the pepper)	" 8,600

ECUADOR.

Ecuador is the Spanish for *Equator*. The Equator passes through this Republic, hence its name.

Ecuador, with Venezuela and New Grenada, once formed the Republic of Colombia. Torn by faction and civil war, they separated about 30 years ago, and set up each a nationality of its own.

The climates and capacity for production are similar to those of the United States of Colombia.

Ecuador, with a population of 1,300,000, of which one-third are Indians, 8,000 negroes, and 37,000 mixed breeds, is fine for fruits and flowers. It is not so large as Texas.

Quito, its capital, having a population of 80,000, is situated 9,528 feet above the level of the sea. It lies at the foot of Pichincha, a volcanic mountain 18,976 feet high, with a crater half a mile deep.

No less than eleven peaks, all white with their snow-caps, are in full view from the *Plaza*, or great public square of this city.

Guayaquil, having a population of 18,000, is the port of Quito, and the principal seaport town of the Republic. From it are exported hides, straw hats, grass hammocks, timber, cacao, Peruvian bark, and tobacco.

It is situated on the Guayaquil river, which is navigable thence to the sea for large vessels. The waters and shores of this stream, where there is no winter to freeze or to chill, are very prolific. The branches which hang in the river are often loaded with oysters, which cling to them as they do to the rocks—hence it is said that in Guayaquil " oysters grow on trees." Iguanos, alligators, and reptiles abound. The natives on the banks of this river build their huts on piles, to prevent their children from being devoured by these monsters.

It is here that you see the bud, the blossom, and the delicious fruit on the orange-tree all at the same time.

There is a famous old University at Cuença ; and Loxa, which is 6,760 feet above the sea, was once famed for its cinchona forests.

The bark of the cinchona-tree is the well-known " Peruvian bark" of commerce, and yields to the pharmacopia of our times the valuable medicine of quinine.

Instead of cutting down the tree, stripping all the bark from the trunk, and leaving the stump to put out a new growth, the custom among these thriftless people was to strip the tree as it stood, as high as they could reach, and then leave it standing with most of its bark still upon it. All trees left in that condition are attacked by a worm that destroys them root and branch.

This tree is indigenous to the eastern slopes of the Andes, from Bolivia to the United States of Colombia, and to no other part of the world.

The cinchona forests are nearly all destroyed.

The hamlet of Antisana, at the height of 13,500 feet, was long thought to be the highest human habitation on the earth ; but there are in Peru inhabited places much higher, such as the village of Tacora, which is 196 feet higher, and the Relay House of Rumihuasi, in Peru, 15,540 feet. This station is situated at the edge of the snow-line.

At the Equator the snow-line is 16,000 feet above the level of the sea. Beyond that height the air is too attenuated for long-continued human existence.

There is about the Equator in this Republic a group of remarkable mountain-peaks.

Among them is the dome-shaped Chimborazo, one of nature's most imposing structures, standing at the enormous elevation of four miles in perpendicular height above the level of the sea. The Andes also have higher peaks than Chimborazo.

Mariners have descried this mountain at the distance of more than one hundred miles out at sea, and by moonlight the author has seen it at the distance of ninety miles.

Last in this wonderful array of burning mountains, with their layers of colored snow, we come to the terrific and awfully grand Cotopaxi, the loftiest volcano of the Andes.

It is near the Equator, and stands in perpendicular height 18,870 feet above the surface of the sea.

In its eruptions, and with a noise that is said to have been heard at the distance of six hundred miles, it shoots out a column of flame into the upper air half a mile high.

All the way from the Straits of Magellan up, there is, arranged along on the tops of the Andes, like a line of sentinels, a succession of these snow-capped volcanoes.

The Galapagos Islands, situated *under* " *the line,*"* belong to Ecuador ; they are the only inhabitable group of islands in the Pacific Ocean that were uninhabited at the time of their discovery.

Questions.—What are the area and population of the United States of Colombia?—In what country does the Panama railway lie?—What do we get from Panama?—What are the chief agricultural staples for which the climates of these States are adapted?—Where are the chief branches of manufacture in South America?—Describe the air volcanoes.

What is the capital of the United States of Colombia?—What is its population?—What is said of Bogota and the ascent of its surrounding peaks?

Describe their climate.—What inconvenience do travellers experience when they ascend high mountains?

What are the mineral resources of the United States of Colombia?

What are the area and population of Venezuela?—What large river has it?—How does the face of the country compare with that of the United States of Colombia?—What are its chief sources of wealth?—What are the Llanos?—Describe the seasons.—The overflowing of the Orinoco.—The aspect of the country in the rainy season, and then in the dry.

Describe the flora of Venezuela, with the mimosa and the "cow-tree."—What does Venezuela export?

What is the population of the capital?—Describe the *silla* of Caracas.—What place in Venezuela is celebrated for its tobacco?—For what is Caracas celebrated?—Tell about the ancient city of Cumana, and the celebrity of the Ciudad Bolivar.

What are the Guyanas?—Describe the Guyanas and tell their size.—What is their population?—Their climate and productions?—Describe the rivers and the rainfall of that part of this country.—Describe the forests and rivers of these countries, with the beasts, birds, fish, and reptiles that are found in them.—What celebrated flowering-plant was first discovered in the Berbice ?—Can you tell anything about the Victoria Regia?—By whom are the Guianas inhabited?—What is the population of their capitals?—What do we get from there?

* The *Equator* is called by sailors "the line."

How does Ecuador derive its name?—What is the port of Quito?—What paradox does the Guayaquil river present?—Which city is situated at the greatest height above the sea, Mexico, Bogota, or Quito?—Describe the situation of Quito.—Describe the situation of Guayaquil.—Tell its exports.—Tell about the river and the country-houses on its banks.—What is the chinchona-tree?—Describe these States in their orographical aspects.—(The scholar may examine, at this point, the Orographic View of the Valley of the Amazon, p. 96).

LESSON L.

Equatorial South America.

BRAZIL.

The chief magistrate of Brazil is the only ruler in America that wears a crown.

This empire lies between the parallels of 4° north latitude and 30° south latitude. For geniality of climate, breadth of border, and capacity for production, it is surpassed by no country on the globe.

With an area (3,230,000 square miles) larger than that of the United States without Alaska, Brazil has a little more than one-fourth as many inhabitants as we have. It is a limited monarchy.

Brazil was accidentally discovered by a Portuguese navigator in the year 1500. He was bound to India, and, much against his will, was drifted to the westward by the trade-wind, and found himself on a lee-shore near Cape St. Roque.

Owing to this circumstance Portugal asserted her rights as a discoverer, and Brazil became a Portuguese possession, and was colonized by Portugal. Its inhabitants are of different races and of mixed bloods.

The "King of Rivers," as the Indians call the Amazon, drains the largest portion of Brazil.

THE LOWER AMAZON.

This river has tributaries that in their course traverse more parallels of latitude than the Mississippi does. The La Plata, too, has its head-waters in this empire. There is a gentleman in the province of Matto Grosso who has in his garden two never-failing springs. One flows northwardly, into the Amazon, the other southwardly, into the La Plata, and the distance between these two rivers is only some three leagues, so that if a canal were cut across this portage, inland navigation would be possible from Buenos Ayres, up the La Plata into the Madeira, thence into the Amazon and the Rio Negro, thence through the Cassiquiare into the Orinoco.

The language of the country is Portuguese, and the religion, like that of all Spanish America, the Roman Catholic.

Its principal industries are agricultural, pastoral, and mining.

Coffee, at present, is the great agricultural staple, and the United States the principal consumer of it. Both Europe and the United States are supplied with this berry mainly by Brazil.

Indigo, sugar, mandioc, and cotton are by no means unimportant articles of cultivation and export.

The cotton of Brazil ranks, in the Liverpool markets, with our own, which is the most esteemed of any.

Cattle, as in the La Plata country, are raised chiefly for their horns and bones, hides and tallow, large quantities of which are brought to the United States.

The richest gold-mines in the world are in Brazil, which also has diamond-mines, and precious stones of rare beauty and great value, with exhaustless treasures of the baser metals, such as copper, zinc, lead, iron, etc.

The diamond-mines are in the province of Minas Geraes.

Besides cattle, the plains and forests of Brazil abound in birds and insects of the most brilliant casques and beautiful plumage.

The catching of them, for ornamental work, is a special branch of industry. The feather-work from the convents of Brazil is famed for its elegance.

From the forests of Brazil we get our chief supplies of india-rubber. The best comes from the Amazon.

Other gums, spices, nuts, perfumes, drugs, balsams, and medicines, such as rhubarb, sarsaparilla, jalap, sassafras, holywood, dragon's blood, licorice, and ginger, amounting to a large sum in value, are likewise exported from Para.

Brazil, considering its extent, is one of the most abundantly watered countries in the world. The number and length of the rivers indicate this.

The waters of the Amazon abound in a species of turtle that is highly esteemed as an article of food, and valued on account of the oil obtained from its eggs. At the laying season it scrapes a hole in the sand in which a single terrapin will deposit a half-bushel or more of eggs; it then fills up the hole with sand and leaves them to hatch by the warmth of the sun. The egg-hunters collect many millions of them annually.

On the banks of the Amazon the boa-constrictor and alligators are abundant, and all along its borders vines and parasites cover the trees and make the woods gay with the most beautiful flowers. Parrots build their nests, and other birds sing among the branches of the trees, which are so closely matted together that the monkeys may travel for days on the tree-tops without ever coming to the ground.

The valley drained by the Amazon contains an area of 1,796,000 square miles.

The Cassiquiare (*cah-see-kee-ah're*) forks, one branch flowing north into the Orinoco, the other south into the Amazon. It unites these two river systems, and brings the valleys drained by them into one hydrographic basin.

If, therefore, we include the valley of the Orinoco, we have, in South America, a river basin that contains more than two millions of square miles.

Belgium has a population of 440 persons to the square mile. According to this rate, there is room in this magnificent river basin for more than 800,000,000 of people.

There is a line of steamers on the Amazon that ply regularly between Para, at its mouth, and Nauta in Peru. (See Map, p. 96.)

The Madeira is its largest tributary. Its navigability is interrupted by rapids at the distance of about 1000 miles from its mouth. Above these falls it is again navigable for five hundred miles or more.

Rio de Janeiro, generally called Rio—the capital of the empire, and its chief city—with a population of 420,000, and a harbor unsurpassed, is the largest city in the Southern Hemisphere; and, with its museums, its institutions of learning, its operas, prados, public promenades, and botanical gardens, is the most splendid capital in the New World.

Bahia, with a population of 150,000 ; Pernambuco, of 120,000; and Para, of 10,000, are the other chief towns in Brazil.

The United States have a large trade with Brazil.

Questions.—Between what parallels of latitude does Brazil lie ?—Describe it.—When and how was it discovered ?—By whom was it settled ?—By whom is it now peopled ?—What is its form of government ?—Describe the climates and productions of Brazil.—Describe the garden springs of Matto Grosso.—Suppose a voyager should undertake to go in a canoe, by inland navigation, from the mouth of the La Plata to the mouth of the Orinoco, what route would he take, and what portage would he make ?—What are the languages, the religion, and the chief industries of Brazil ?—What are its staple productions ?—What do we get from there ?—Where are the richest gold-mines in the world ?—What other metals and precious stones are found there ? —Where are the diamond-mines ?—Name the principal articles that we get from Brazil.—What reptiles do you find in this warm and moist country ?— Describe the forests.—How large is the Valley of the Amazon ?—What river unites it with the Orinoco ?—How large is the hydrographic basin thus formed ?—Suppose it to be as thickly settled as Belgium, what would be its population ?—How far is the Amazon navigated by steamers ?—What is its largest tributary ?—Describe the face of this country.—What is the capital of Brazil ?—Describe the harbor of Rio.—Name some of the chief towns, with their population.—Have we much trade with Brazil ?

LESSON LI.

The Andean States of South America. (See Map, p. 102.)

Ecuador, like the United States of Colombia and Peru, lies on both sides of the Andes. Each of the three contains, as Bolivia does, a large area of table-land, which is dotted with volcanoes and snow-capped mountains. The base of these mountains rests on this table-land, which is itself from 6,000 to 12,000 feet above the sea.

Of course, then, all four of these countries have every variety of climate that can be found in other parts of the world, between the regions of eternal frost and everlasting spring.

In our country these climates are spread out horizontally, and extend along the seaboard all the way from the Gulf of Mexico to the icy regions of the north.

STREET IN RIO.

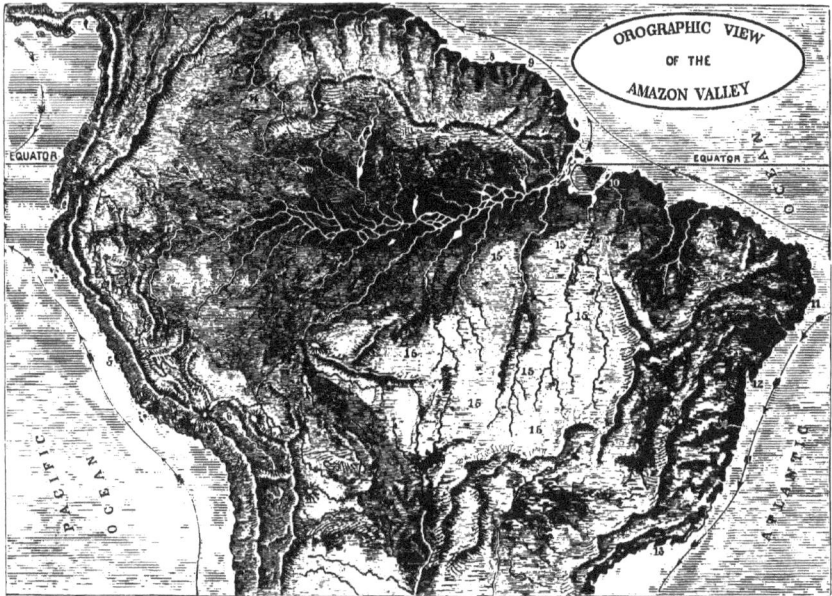

OROGRAPHIC VIEW OF THE AMAZON VALLEY

In these four countries, they are not stretched out, but piled up one above the other.

People here who want to go to a warm climate are sure to find it in the plains below; those who desire a more bracing air have but to climb the mountain-sides a little way, and they find it.

The Amazon rises in Lake Lauricocha, on the top of the Andes. It, with its tributaries, the Huallaga (*hwal-yah'gah*) and the Ucayale (*oo-ki-ah'lā*), rises each in a valley of its own; they then take a northerly course for several hundred miles before they come to a gap in the Cordilleras; here they break into falls and rapids which obstruct navigation. They then unite into a gentle but mighty stream, fertilizing the plains below and conveying to the sea food for its inhabitants.

The Ucayale, with its windings, is said to be navigable for 1,200 miles above its junction with the Amazon, thus showing how gently this mountain-valley, which is drained by it, slopes toward the north.

These three streams show that the Andean plateau is laid off by nature into a series of parterres and terraces, each containing areas of hundreds of square miles, and rejoicing in a climate singularly equable.

[The following is a list of names corresponding to their respective numbers, as found on the Orographic view of the Amazon Valley:

1. Bogota.	7. Cuzco.	13. Rio.
2. Quito.	8. Paramaribo.	14. Llanos of Orinoco.
3. Chimborazo.	9. Cayenne.	15. Table-lands of Brazil.
4. Lake Lauricocha.	10. Para.	16. Itambe.
5. Lima.	11. Pernambuco.	17. Nauta.
6. Lake Titicaca.	12. Bahia.	

Arrows show ocean currents.

NOTE.—Let the pupil be required to pause and without the aid of the numbers, point out each and every place indicated in the above view. Especially let him endeavor to ascertain the gradation of ascent he would make in going from the mouth of the Amazon to Lakes Lauricocha and Titicaca.

A clear understanding of the orography of a country is a prerequisite to any knowledge of its general geography, as also to the first attempts at Map Drawing.]

Thus, a traveller ascending the Amazon would leave the hot climate of the main stream, and find in the valley of the Ucayale a spring-climate all the year long. Climbing the next step and passing on into the valley of the Huallaga, a milder climate still would await him. Ascending thence into the valley of this Upper Amazon proper, he would find himself in a superb wheat, corn, cattle, hemp, and tobacco country, with cool nights, rare frost, bright skies, and pleasant days at all seasons.

In ascending the Amazon from its mouth, you meet with no falls or rapids until you pass Nauta, in Peru.

From latitude 30° north to latitude 30° south the prevailing direction of the winds all around the world

ıese parallels the trade-winds
, they are from the northeast ;
are from the southeast.

ın will see that, owing to the
of South America, the north-
s come from the sea strike the
also do the southeast trades.
ın they reach the land, being
ing with moisture. As they
they ascend the great Ama-
ısit their moisture.

· that all the way from the sea up to
try. (See Orographic View, p. 96.)

ıend, get cooler and cooler,
down their moisture in the
w, and finally reaching the
lleras, they pass over to the
·obbed of moisture. Every
ıng from them while crossing
·eturning back eastwardly to
mountain streams, winding
ourses.

s dry winds, have no moisture for the
rainless, and quite different from the
aspects.

ıthed with trees and verdure.
are parched and dry : they
s the rock.

en the Andes and the shores of the
reams, formed by the melting snows of
rinking and for irrigating their fields.
· by irrigation.

Peru is, wherever there is
without glass. It produces
ıud delicious fruits, and has
l comfort is unsurpassed.

ater, every plant, fruit, and flower that
in the greenhouse flourishes here.

ı be planted only once in
in six. They produce con-

stalks are perennial ; they
ll the year, and the cotton-
f the streams like bushes in

physical relations be.ween
-lines and winds.

Creator on the east instead of along the
l South America to the westward of

; east and west, instead of north and

south, there would have been no Amazon ; and the physical aspects of this
beautiful country would have been quite different from what they are.

Oysters abound in the Guayaquil (*gwi-ah-keel*) river ; but south of that the
situation of the Andes with regard to the winds and the sea-shore is such as
to deny the sea fresh water enough to make brackish bays, and creeks where
the oyster can live. Therefore the fish markets of Peru, Bolivia, and northern
Chili are without oysters.

South of latitude 40° the prevailing winds are from the
west, and there the condition of things is reversed ; there
the western side is the rainy,and the eastern the dry side.

PERU.

Peru is more than twice the size of Texas. It has an
area of 510,000 square miles, and a population of
2,500,000.

All of Peru west of the Andes is rainless.

The sea along this rainless part of the coast is also
the most gentle part of the ocean. It is never ruffled by
a storm. Therefore remember *that rainless shores* are
washed by *stormless seas.*

The mines of Potosi and Pasco, with others in Peru,
yielded in colonial times a fabulous amount of silver.

Silver was then used there as the baser metals are with us : tires of car-
riage-wheels, horseshoes, and the commonest household utensils were of
solid silver.

I have seen there, in the early days of independence, clouted Indians
sitting at dinner on the dirt floor of their hut, around a massive silver dish,
all dipping and eating without the aid of knife, fork, or spoon.

The Andes of Peru are exceedingly rich in silver,
quicksilver, copper, lead, and iron.

Guano, in the abundance and quality of which Peru surpasses all the
world, has recently come extensively into use as an agricultural manure.
The Chincha Islands, which lie in sight of the coast, are covered with it.
Nature has piled it up there in a merchantable form, all ready for market.

The government now derives a clear revenue of some $10,000,000 or
$12,000,000 annually from the sale of this fertilizer, and the quantity con-
tained on these islands is worth, at present prices, nearly $400,000,000.

On the Andes of Peru and Bolivia are found in
great numbers what is called the "Peruvian sheep."
It is not really a sheep, but a species of camel. It consists of four varieties, the llama, the alpaca, the vicugna
(*vǐ-koōn'yah*), and the guanaco.

The exports of Peru consist first, of guano, which is
the largest and most valuable item, to which may be
added the wool and hair of the "Peruvian sheep ;" bul-
lion, cotton of excellent staple ; a few gums and drugs;
a little sugar, cocoa, coffee.

Lima, the capital, has a fine library, a noble cathedral,
and 54 churches and convents, all Roman Catholic. It
was founded by Pizarro, in 1534, and contains his remains. It is seven miles from Callao, its port.

CATHEDRAL OF LIMA.

Lima is a fine old Spanish city, with a population of 120,000.

Cuzco, with its 47,000 inhabitants, renowned for its Temple of the Sun, and its ancient glory as the capital of the Incas, is the second city of importance ; and Arequipa (*ar-ē-ke'pah*), with a population of 20,000, is the next.

AREQUIPA IN 1868.

The great volcano of Arequipa, a truncated cone, rises on the outskirts of the city to the height of 20,320 feet. In August, 1868, a fearful earthquake visited the City of Arequipa, and nearly destroyed it.

Pasco is celebrated for its silver mines; Pisco for its grapes and whisky; and Huasco for its white and transparent raisins.

The inhabitants of Peru, like those of most of the Spanish American Republics and Brazil, consist chiefly of Indians and mixed breeds.

BOLIVIA.

The physical geography of Peru is repeated in Bolivia, which, with a population of 1,987,000, and an area of 536,000 square miles, is six times the size of Great Britain. (Refer to Orographic View, p. 96.)

It was named in honor of the South American "Liberator," Bolivar.

Most of Bolivia that lies on the eastern slopes of the Andes is a wilderness. It is above the falls of the rivers that are tributary to the Amazon and the La Plata, and of course is cut off from the Atlantic. The climates and productions of this part of Bolivia may be said to include those of all the habitable portions of the globe.

Here, one seated at the foot of a mountain, and surrounded with the luscious fruits of the tropics, may, casting his eyes up toward the snow-capped peak above him, take in at one view the whole range of the vegetable kingdom.

You can see, by looking at the map, that Bolivia has only a narrow strip of sea-coast, which practically is of no value. It is at the head of the great Desert of Atacama (*a-ta-ca'mah*), and forms a part of it.

To avoid the port-dues and custom-house charges of Peru, Bolivia resorted to the expedient of making Cobija, which is an open roadstead, a free port.

A free port is a port where there is no custom-house, and no duty to pay.

Coca is produced in Bolivia and Peru, and is found nowhere else.

This plant is used by the natives somewhat as the betel-nut is used in the East Indies, and as tobacco is used with us; but it is said to possess virtues which are not claimed for either of the others.

The leaf is chewed with unslaked lime, as the betel is, and those who use it can go several days without fatigue, thirst, hunger, or sleep.

In Cochabamba, we find a number of manufactories of cotton fabrics and glass-ware. In La Paz, the largest city in the Republic, there are manufactories of hats and woollens from the fleece of the Peruvian sheep.

Sucre, the capital, standing 9,300 feet above the sea, has a splendid cathedral, a school of mines, and excellent colleges.

Potosi, in its palmy days, when the mines in its vicinity were so profitably worked, numbered 130,000 inhabitants. It is 13,314 feet above the sea.

No city was ever before built at an elevation so high. Since the mines have ceased to be worked it has sadly fallen off, and now it is estimated to contain not more than 20,000 inhabitants.

Lake Titicaca reposes, with its brackish waters, at the height of 13,000 feet above the level of the sea.

Here the atmosphere is so attenuated and its pressure so diminished, that the lifting-pump, which will raise water from the wells along the sea-shore to the height of 32 feet, will not raise it from this lake more than 20 feet high.

The temperature at which water boils is used to determine the height of mountains. The higher you go the more easily water boils; so that on the top of the highest mountains, water that is *boiling*-hot is not much more than milk-warm.

Under this diminished pressure evaporation is enormously active. It takes up and carries off the water from the lake as fast as the rivers pour it in.

The lake, with its islets, its shores, and surroundings, is a picture of silver, embossed with emeralds, and set in a mountain-frame tipped with frosted work.

Lake Titicaca is navigable by large vessels, and it is subject, as all these high table-lands are, to furious storms of wind and rain, snow, thunder, and lightning.

NATIVES NAVIGATING LAKE TITICACA.

Among the rugged heights of the Bolivian Andes the condor builds his eyrie. He is the largest bird of flight in the world, and for strength of wing and force of beak no other can come near him. He can carry off young calves, sheep, and goats in his talons.

CHILI.

Chili, with a population of 2,085,000 souls, and distributed over an area of 133,000 square miles, is about half the size of Texas.

It is a narrow strip of country lying between the Andes and the sea. It is beyond the Tropic of Capricorn, but its northern borders come within the rainless region. Further to the south it reaches those latitudes where, during certain seasons of the year, the west winds are the prevailing winds —where the land receiving these winds fresh from the sea wrings their moisture from them. They then pass over upon the plains of Patagonia as winds without rain.

In consequence of this change in the direction of the dominant winds, vegetation and sterility change sides. In Bolivia and Peru the western slopes of the Andes are barren —the eastern clothed with verdure; but in Southern Chili the western slopes are ever green, and their fertility makes Chili the granary of all South America on the Pacific Ocean.

The wheat in Chili has the largest grains I have ever seen, and I have seen ship-loads of it piled up on the wharves of Callao, lying there for months at a time in the open air, with no more protection from the weather in that rainless port, than if it had been a pile of paving-stones.

Chili is a fine cattle country. The climate of Chili is the duplicate of that of California. The two countries are equidistant from the Equator, but on opposite sides of it; consequently the seasons are opposed, for when it is the rainy season in one, it is the dry season in the other, and the winter of one is the summer of the other.

Chili and Paraguay have been less afflicted with revolutions and civil wars than any of the other Spanish American Republics. Within their borders, industry is less timid and more energetic than in any other of their sister Republics. No nation on this continent enjoys better credit in the money-markets of the world than Chili.

A railway is in process of construction from Chili across the Andes, to connect with the Argentine railway, its eastern terminus is in the City of Buenos Ayres.

Chili is also a mining country. Its copper and silver mines are energetically and profitably worked.

Between Patagonia and Panama there is not a single river emptying into the Pacific ocean that is navigable for more than a few leagues from the sea. The mountains are too near the coast, and the watersheds are too steep to allow the drainage to gather into large streams. Aconcagua, an extinct volcano, 23,944 feet above the sea-level, is the highest mountain of the New World.

Juan Fernandez, the scene of Robinson Crusoe's adventures, is an island in the Pacific belonging to Chili.

The northern end of Chili lies on the dry side of the Andes; it is included in the desert of Atacama.

About thirty years ago, an Indian, being benighted in this waste, gathered up such materials as he could find and built his fire against a rock, as he thought. In the morning he found the back of his fireplace all silver, which proved to be worth not less than $20,000. This led to the development of the mines in the vicinity of Caldera.

Chili, like Mexico, and most of the Spanish American Republics, has good common schools, that are supported out of the public treasury, and to which all who will may send their children.

Santiago, the capital, with a population of 115,000, situated 90 miles inland from Valparaiso, is the most elegant capital in Spanish America. Its society is refined. The city is situated among the Cordilleras of the Andes.

"Valparaiso," the *Vale of Paradise*, the port of Santiago, is delightfully situated on the sea-shore. It is the largest and most flourishing seaport along the whole coast of Spanish America, from Patagonia up to the Gulf of California. It has a population of 80,000.

Valparaiso has a large trade in wheat, wool, hides, and copper.

PATAGONIA AND THE FALKLAND ISLANDS.

Patagonia, with *Tierra del Fuego*, is thinly peopled. Their inhabitants are savages of a low order.

Tierra del Fuego and Cape Horn—the *land of fire*, and the *furnace cape*, so called because of the fiery furnaces (ornos) and burning volcanoes which the

Spanish discoverers fancied they saw there—are both islands, the former being separated from Patagonia by the Strait of Magellan (*ma-jel'lan*), which bears its name in memory of the bold old navigator who first discovered it. His name has also been carried up among the stars, for he was also the first to observe, and to call the attention of astronomers to those curious nebulous clusters in the sky—the Magellanic clouds—which, in those latitudes, are almost directly overhead. (See Map Studies of Africa.)

To avoid the dangers of Cape Horn, small vessels and steamers generally pass through the Strait of Magellan. Its shores are bold, its waters deep, the rise and fall of the tide considerable, the currents are strong, the winds baffling, and the weather squally. In places the water is matted over with *kelp*, a long and cord-like seaweed, which is liable to foul the rudder or disable the propeller.

Nevertheless the prudent navigator, with steam to help, prefers to take his chances through this Strait rather than encounter the fierce winds and heavy seas off Cape Horn. The mercury in the barometer, even in fine weather, off Cape Horn, is remarkably low.

The Falkland Islands belong to Great Britain, and contain a settlement of not more than 1,000 persons. Vessels doubling Cape Horn, both coming and going, must pass them.

The geographical position of these islands, therefore, makes them, in the hands of a great naval power, a military outpost of importance.

They are almost treeless, but their tall and picturesque tussock-grass is sometimes mistaken by mariners, as they sail along the shores, for palmetto groves.

THE FALKLAND ISLANDS.

They are a favorite resort for the penguin, albatross, cape-pigeon, and other sea-birds which go there to lay, hatch, and breed.

The albatross of Cape Horn, though but little larger than a goose as he sits the water, yet of all the aerial tribe unfolds the greatest *spread* of wing. I have seen one taken there that measured 16 feet from tip to tip.

Questions.—Describe the table-land of the Andes.—(See Orographic View.)—Describe the climates of the table-land.—Are they spread out there as they are here, or are they piled up?—Where does the Amazon rise?—By what two tributaries is it first joined?—Describe the valleys drained by them.—To what town is the Amazon navigable?—What crops thrive on the table-land?—How far, by the map, would you judge it to be from Nanta to the mouth of the Amazon?—Between what parallels and from what quarter do the trade-winds blow?—What are the winds that bring the rains to South America?—What part of it lies within the trade-wind region?—Can you explain why the trade-winds bring no rain for the western slopes of the Andes?—Where does their water come from?—Describe the climate, the vegetables, the fruits and flowers there.—Suppose the Andes had been placed near the east instead of the west coast, what would be the consequence?—What is the vege-

tation along the shores of Peru?—South of latitude 40°, which is the rainy side of the Andes?

Describe the rainless region, the mines, and the wealth of Peru.—What of guano?—The Peruvian sheep?—The exports?—Lima?

Which is the largest, Peru or Bolivia?—Whence the name of Bolivia?—Describe its situation with regard to commercial outlets.—What is coca?

Name the chief towns of Bolivia.—Their elevation above the sea.—Why does water at great heights boil before it gets hot?—What advantage is taken of this circumstance?—Describe Lake Titicaca.

Name the largest seaport town on the west coast of South America.—Why are there no large rivers between Patagonia and Panama?—What is the highest mountain in America?—How high is it?—To whom does Juan Fernandez belong?—Can you repeat the law about the snow-line?—Are there any common-schools in Chili?—What part is desert?—What can you tell about the climate of Chili?—Compare it with that of California.

Describe Patagonia, Tierra del Fuego, and Cape Horn.—Why is the doubling of Cape Horn so difficult?—To whom do the Falkland Islands belong?—In what aspect do they have importance?—What is said of the Cape Horn albatross?

LESSON LII.

And the Republics of Uruguay and Paraguay, all lie in the valley of the Rio de la Plata, and their physical aspects are the same. Like Venezuela, they are agricultural and pastoral.

What we call our western plains and prairies are there called "pampas." With occasional strips of woodland in sheltered places, the pampas reach from the sea to the mountains, and support enormous herds of wild cattle.

At the time of its discovery, South America had no animals answering to our elk and buffalo; neither had we any birds answering to their ostrich and condor.

The wild cattle of the pampas are chiefly descended from cattle brought from Europe. They became so numerous that immense herds of mares were slaughtered solely to obtain grease for the soap-boiler.

Instances have been known in which, owing to the scarcity of wood, carcasses of sheep were used for the brick-kilns of Buenos Ayres.

Great numbers of cattle are still slaughtered for their hides, hair, horns, and bones, which are brought to our country to be dressed and prepared for use.

This country is at present attracting more immigration than any other part of Spanish America, and its government has decided upon a liberal policy to encourage it, exempting the immigrant from military conscription and his farming implements from custom-house duty, at the same time making liberal concessions with regard to lands. Thus encouraged, a tide of immigration is setting thitherward from Italy, France, England, and the United States.

The La Plata with its valley is the Mississippi of South America. Its course, unlike that of the Amazon, is not along parallels of latitude, but like that of the Mississippi, across them. With every bend in the river you reach a different latitude, and with every new latitude there is a change of climate, and with change of climate there always follows change of human wants and change of productions.

The course of the La Plata, with its tributaries, from north to south traverses 23° of latitude.

The valleys drained by these two rivers embrace these areas:—the Mississippi, 982,000 square miles, and the La Plata, 886,000 square miles. The La Plata crosses more degrees of latitude; the Mississippi drains the broadest but the La Plata the longest valley; and within this long valley are soils adapted to the cultivation of wheat, corn, coffee, tobacco, sugar, rice, cocoa, hemp, flax, indigo, and mandioca.

Land is cheap and abundant, and the pampas for pasturage are boundless commons.

There is a large exportation of horns, hair, hides, jerked beef, etc., from this country to Europe.

The climate of this river-basin is free from frost and remarkably healthy. Furious tornadoes, called *pamperos*, sweep across the pampas to the sea. They are sometimes accompanied by those fearful discharges of thunder and lightning which engender the "fulgurite." In the absence of pinnacles the lightning often strikes the ground. It then makes a hole in the earth, melts the sand, and leaves around the hole a vitrified funnel-shaped mass.

As in all the other parts of Spanish America, the Spaniards, though in the minority as regards numbers, are in the ascendant as to control and management. Theirs is the language of the country. The majority of the inhabitants consists of Indians and cross-breeds, and is Roman Catholic in religion. Those that live on the pampas are most expert horsemen. They have the habit, when going to the charge in battle, of throwing the heel across the horse's back and riding under his belly. These are the *Gauchos*. They are very dexterous in using the lasso. Armed only with this, they chase over the pampas the ostrich, the wild horse, and the bullock, and they throw the lasso with such precision that they can catch the bird or the beast by the foot while it is yet lifted in flight.

GAUCHOS CATCHING CATTLE.

The governments of all the fifteen nations of Spanish America are republican in form.

THE ARGENTINE CONFEDERATION.

This Republic, sometimes called Argentina, contains only 1,465,000 inhabitants, though it has an area of 826,000 square miles. It has a larger area than that of any nation in Europe except Russia.

Buenos Ayres, the capital, is also the chief town of the confederation, and has a population variously estimated at from 120,000 to 200,000.

You observe that the La Plata has no delta. Its mouth is an estuary of the sea, which as high up as Buenos Ayres is still 30 miles broad. The water as you approach the shore is shallow, so that vessels have to anchor several miles out and use lighters for loading and unloading.

All this part of the country is stoneless and treeless.

There is a line of railway in process of construction by way of Rosario and Cordova, which is designed, at no distant day, to connect Buenos Ayres, on the Atlantic, with Valparaiso, on the Pacific.

The Salado is remarkable for its brackish water.

Tucuman, with its 11,000 inhabitants, stands in the "Garden of Argentine;" San Juan, at the foot of the Andes, is a flourishing new town of 20,000 people.

The first congress of the La Plata States was held there in 1816.

PARAGUAY

Has an area of 126,000 square miles, and a population of 1,337,000.

Though called a Republic, Paraguay has been governed by three dictators ever since the loss by Spain of these colonies. First by Doctor Francia who would allow no foreigner to come within his dominions; then by Lopez, and afterward by his son; and of all the Spanish American countries, it has had the most stable government and prosperous industry. Paraguay produces a holly, which the inhabitants call *yerba* (*the vegetable*), out of which maté is made. Maté is a tea, and is a favorite beverage in Brazil and throughout all the La Plata country, Bolivia, Chili, and Peru.

It is the only nation in America without a sea front.

Asuncion, the capital, is also the chief town. It is estimated to contain between 12,000 and 20,000 inhabitants.

The La Plata, with the Parana and Paraguay, is navigable for steamboats far above Asuncion.

URUGUAY

Is the smallest of the South American Republics, and is not quite as large as Missouri. It contains a population of only 241,000.

Its capital, Montevideo, with a population of 100,000, is situated near the mouth of the river, where it is 60 miles wide, and has an extensive commerce both with England and the United States. Lines of steamers now ply regularly up and down the La Plata.

Questions.—Where is the La Plata country?—By what States is it occupied?—What are the chief industries of these States?—Describe the pampas.—Is there any immigration to the La Plata country?—Why is the La Plata called the Mississippi of the southern hemisphere?—Compare the valleys of the two rivers, their length, and the direction in which they run.—Describe the climate, productions, and exports of the La Plata country.—Its language and religion.—What are *fulgurites?*—Pamperos?—Who are the Gauchos?—What is the form of government?—What is the area and population of the Argentine Confederation?—Of Uruguay?—Of Paraguay?—Describe the Argentine Confederation.—Its capital and chief towns.

Describe Tucuman, and the route of the proposed railway from Valparaiso to Buenos Ayres.—How has Paraguay been governed?—What are its area and population?—What is maté, and where does it come from?—What is the population of Uruguay?

SOUTH AMERICA

Statute Miles

LESSON LIII.

Studies on the Map of South America.

Boundaries.

How is South America bounded?—Through, and near what countries does the Equator pass?—Between what meridians of longitude does South America lie?—What parts of South America are traversed by the Andes?—By the Amazon river?—By the Rio de La Plata?—What countries lie south of the Tropic of Capricorn?—Upon what parallel of latitude does Cape Horn lie?—Through how many degrees of latitude does South America extend?

How do you bound the United States of Colombia?—How do you bound Venezuela?—The Guyanas?—Ecuador?—Brazil?—Peru?—Bolivia?—Chili?—The Argentine Confederation?—Paraguay?—Uruguay?

Islands.

Where are the Falkland Islands?—Chiloe?—Juan Fernandez?—The Chincha Islands?—St. Felix?—Barbadoes?—Staten?—Tierra del Fuego?

Mountains.

Name the chief range of mountains.—In what direction do they run?—What mountains can you name in the northeastern part of South America?—What, in Brazil?—The highest peak of the Andes?—Where is Pichincha?—Cotopaxi?—Chimborazo?—Aconcagua?—Illimani?

Rivers.

What is the course of the Magdalena river?—Of the Amazon river?—Of the Orinoco?—Of the Rio de La Plata?—Of the Madeira?—The Ucayale?—The Huallaga?—The Tocantins?—The Cassiquiare?—The Parana?—The Para?—The Paraguay?—The Uruguay?—Does the Amazon receive most of its tributaries from the south or from the north?—Name the largest tributaries of the Amazon Of the La Plata.—Where does the Magdalena river empty?—Into what ocean is South America chiefly drained?—Where is the Purus?—The Pilcomayo?—The Mamore?—Tell where the Essequibo river is.

Gulfs and Bays.

Begin at the Isthmus of Panama, and name the chief bays that indent the coast of South America.—Name, in the same way, the chief gulfs.

Capes.

What are the chief capes of South America?—Name them, in order, from Panama around the continent?—Where is Cape St. Maria?—Cape St. Roque?—Cape Frio?—Cape San Antonio?—Cape Horn?—Cape Blanco?

Lakes.

Name the principal lakes of South America.—Where is Lake Titicaca?—Where is Lake Aullagas?—Guanacache?—Maracaybo?

Political Divisions.

Name the political divisions of South America.—Which is the largest?—Which is the smallest?—How far, in an easterly direction, do the United

States of Colombia extend?—What two important cities on the isthmus, connected by railway, lie within these States?—Does the greater or less part of Ecuador lie east of the Andes?

Chief Cities.

Where is Aspinwall?—Panama?—Bogota?—Caracas?—Georgetown?—Quito?—Paramaribo?—Cayenne?—Rio de Janeiro?—Montevideo?—Buenos Ayres?—Sucre?—Asuncion?—Lima?—Callao?—Santiago?—Valparaiso?—Where is Maracaybo?—Truxillo?—La Guayra?—Cumana?—Cuzco?—Arequipa?—Guayaquil?—Conception?—Villa Rica?—Nauta?—Para?—Maranham?—Pernambuco?—Bahia?—Diamantino?—Cuyaba?—Victoria?—Caldera?—Cobija?—La Paz?—Where is Coquimbo?—Copiapo?—Potosi?—Ayacucho.

Routes and Distances.

In what direction and what distance (always, of course, using the scale of miles) is Bogota from Cartagena?—From Panama?—From Quito?

How far is Quito from Lima?—From Guayaquil?—From Valparaiso?

How far is Quito from the mouth of the Amazon?—How wide is South America on the Tropic of Capricorn?

How far is it from Rio to Asuncion?—To Montevideo?—How far is Asuncion from the sea?—How far, from Rio to Bahia?—To Para?—How far are the Galapagos Islands (Map, p. 20) from the coast of the continent?

Miscellaneous.

Which of the countries of South America has the greatest extent of sea-coast?—What extent of sea-coast has Ecuador?—Can you point out on the map where the chinchona-tree grows?—Can you point out on the map the famous old road of the Incas (see page 90)?—Can you find the hamlet of Añtisana (13,455 feet above the sea)?—Point out the Cumbre Pass. *It leads from Valparaiso to Mendoza, under the shadow of Aconcagua. The summit of Aconcagua is several thousand feet above the snow-line.*

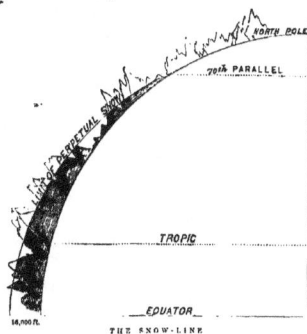

At the Equator the snow-line is 16,000 feet high, and the further you go from the Equator, either to the north or the south, the lower is the snow-line, until, about the 70th parallel of latitude, it touches the earth's surface. This is a geographical law which you ought to remember.

Where is Hermit Island?—Trinidad?—The desert of Atacama?

SCENERY AND ANIMAL LIFE IN EUROPE.

LESSON LIV.

Europe. (Map, p. 121.)

1. Political Geography.—Europe is an old country, and its nationalities count their ages by centuries.

The States of Europe are four Republics, four Empires, and fourteen Kingdoms, besides a number of Duchies, Principalities, and free States which belong chiefly to Germany, and are not recognized among nations as separate and independent powers.

The new German Empire includes the kingdoms of Prussia, Saxony, Bavaria, and Würtemberg—all of which are one POWER.

Sweden and Norway are under one king, and the Emperor of Austria is king of Hungary also. These four countries, therefore, make but two powers.

Of the nations of Europe one is infidel, and the rest are either Roman Catholic, Greek, or Protestant.

Europe is well developed and overflowing with population.

Two of the States of America are each nearly as large as Europe. But Europe contains seven times the population of the United States, and more than twenty times that of Brazil.

Europe lies chiefly in the North Temperate Zone. The habitable portions of America lie in the Torrid Zone as well as in the two Temperate Zones.

In Europe, land is dear and labor abundant. In America, land is cheap and labor scarce.

From such points of resemblance and contrast between these two countries, we arrive at conclusions of high import to the political geographer; for the facts just stated show that it is easier for the working-man to make a living in a new country, as America, where land is cheap and labor dear, than it is in an old one like England, where land is dear and labor cheap.

Hence the great migration of men from the Old World to the New.

2. Social Features.—To an American who visits Europe for the first time, the most striking features in its political geography are the high state of improvement of the country, the absence of fences, the vast extent of cultivated or improved lands in proportion to woodlands, the number of villages and lordly mansions and spacious barns and outhouses, which such an extent of highly cultivated fields suggests. He is surprised also, at the number of female laborers that he sees in the fields, especially on the continent.

In America, particularly in the Southern States, both farmers and laborers generally reside on the farms ; but in England, France, and Germany they usually reside in towns and villages.

The excellence of the country roads, the size of the carts, and the immense loads that he meets, drawn by one or two horses, also attract the attention of the American in Europe.

3. Population.—Europe is so thickly settled that there is one person to every eight and one-half acres of land, whereas in America there is one only for every one hundred and twenty acres.

4. Occupations.—You may observe by the map that Austria and Prussia have a small extent of sea-coast; consequently, they have never had a large seafaring population. It is the seafaring population of every country that furnishes it with sailors for its navy. Hence Austria and Prussia, though ranking as first-class powers, have never been ranked among the great naval powers of Europe. On the other hand, Greece, Italy, France, Spain, Portugal, England, Holland, and Denmark are on the sea, and are regarded as the maritime States of Europe.

Spain and Portugal are called the *Peninsula.*

5. Natural Peculiarities.—The Volga is the largest, but the Danube and the Rhine are the two most important rivers in Europe, and the Alps the highest mountains. The highest peaks of the Alps are about 15,000 feet above the sea-level.

Do you remember that you have been told to consider, in your geographical studies, the slopes of the mountains and hills in every country as watersheds, and the rivers as gutters for carrying off the water and emptying it into the sea? Now you can tell why, in Europe, we have no such rivers as the Amazon, La Plata, the Mississippi, and the La Plata. You observe that the continent of Europe is not one-third the size of America. The mountain-ranges of Europe lie, some east, some west, some north and south, and some obliquely to these: they therefore divide it out into a great number of river-basins or watersheds, each of which (as you will see by studying the Orographic View of Central Europe, p. 119), is drained directly into the sea.

In North America there is but one grand range of mountains—in North America but two, one of which, the Alleghanies, runs to the northeast, and the other, the Rocky Mountains, to the northwest, with a great valley and immense watersheds between them.

The highest peaks of the Alps are always covered with snow.

Immense fields of ice, called Glaciers, are formed on the sides of the mountains, and are always sliding from the top toward the bottom. They bear to the mountains very much the same relation that snow-slides do to the roof of a house. There are about 400 glaciers between Mont Blanc and the Tyrol in Germany.

6. Climates and Productions.—Stretching from the heated waters of the Mediterranean up to the frozen ocean of the north, Europe has every variety of climates and productions except those of equatorial lands.

14

7. Nationalities.—Only one-third of the Russian and Turkish empires lie in Europe; they are, nevertheless, classed among the European powers.

The European powers, in the order of their respective population, consist of—

Four Empires.—Russia, Germany, the Austro-Hungarian Empire, and Turkey.

Fourteen Kingdoms.—Great Britain, Prussia, Italy, Spain, Holland, Sweden and Norway, Denmark, Portugal, Belgium, Bavaria, Saxony, Würtemberg, Greece, and Hungary.

Four Republics.—France, Switzerland, Andorra, and San Marino.

Five Grand Duchies, Eight Duchies, Four Free Cities, Nine Principalities, One Landgraviate, One Electorate.

All that part of Europe that extends from the North Sea, Denmark, and the Baltic, to Switzerland, Italy, and the Adriatic on the South; from Belgium, Holland, and France, to Russia, Galicia, and Hungary on the east, is called *Germany,* not because it is one power, but because it is inhabited by the "yellow-haired" races who speak German, and who have agreed to maintain a certain community of interests. These people, numbering at least 57,000,000, occupy an area of 332,000 sq. miles in the heart of Europe.

But this country, as the map shows, lacks sea-front, except upon what are called "closed seas," and therefore, though a mighty nation and powerful on land, the Germans have never been ranked as a naval power.

Questions.—1. Is Europe an old country?—How many nations are there in Europe?—How many kingdoms are there in Europe?—Empires?—Republics?—Which countries have the same king?—What is the religion of Europe?—Which is the most densely populated country, Europe or America?—What two American nations are each nearly as large as Europe?—Can you explain why there is such a large tide of immigration flowing from Europe into America?

2. When a traveller from America visits Europe for the first time, what geographical subjects most excite his notice?—Do the farmers and planters in the South generally reside in town or country?—Where do those of Europe reside?

3. How many acres of land in Europe to the inhabitant?—How many in America?

4. Why cannot Austria and Prussia boast of a large seafaring population?—Are they first-class powers?—Have they never been ranked among the great naval powers?—Which are the maritime powers of Europe?—What nations occupy the Peninsula?

5. What are the most important rivers and the highest mountains of Europe?—How high are the tallest peaks of the Alps?—Why do you see no such rivers in Europe as the Amazon, La Plata, etc.?—What are glaciers?

6. What is said of the climates and productions of Europe?

7. Are Russia and Turkey classed with the European powers?—How much of their territory lies in Europe?—Name the Empires of Europe in the order of their population.—Name the Kingdoms.—Which is the largest in size?—Name the Republics, and point them out on the map.—How many Free Cities are there?—How many Principalities?—What nations constitute Germany?—Why are they called Germany?—What is their population?—Area?

LESSON LV.

The United Kingdom of Great Britain and Ireland.

1. Political Geography.—In olden times, England, Ireland, and Scotland were separate and independent kingdoms.

About 250 years ago they were brought together under the rule of King James I., and were called the United Kingdom of Great Britain and Ireland.

The United Kingdom consists of two principal islands and a number of smaller ones in the adjacent waters.

The island of Great Britain includes England, Scotland, and Wales.

Ireland and Scotland are in size each equal to Maine ; Wales to Massachusetts, and England to Alabama.

The government is monarchical, but liberal. The majority of the people profess the Protestant religion, and the Episcopal church is the Church of the State.

This small country has ruled the seas and spread her name in all parts of the world. She has carried her conquests and established her colonies so widely that the sun never sets upon them. Her *colonial* possessions have an area of 8,500,000 sq. ms. and a population of 165,000,000. She rules one-seventh of all the people, and owns one-sixth of all the land in the world. She is the richest nation, and excels all others in the extent of her manufacturing, seafaring, and commercial industries.

We and the English have the same literature and speak the same language. We derive our notions of law and liberty from the same source.

We carry on more commerce with the English and are more closely allied to them than to any other people or nation, and therefore, to an American, the geography of England is almost as interesting and as important as the geography of his own country.

Moreover, it is instructive for you to examine into the geographical conditions that are peculiar to England, and which have helped her to ascend so high in the scale of national greatness and renown.

2. Early and Subsequent Development.—Four hundred years ago, neither the existence of America nor the passage round the Cape of Good Hope was known. In the ignorance then existing as to the science of navigation, commerce was carried on chiefly by caravans overland. Ships dared not launch out upon the broad ocean because they could not find their way ; the instruments of navigation were so rude that voyages were confined to closed sheets of water like the Black and Mediterranean Seas.

In this state of things England found herself excluded, in a great measure, from the commercial circle of the world, which at that time consisted of Europe, Asia, and Africa. Trade was carried on chiefly by caravans.

3. Position.—But after the discoveries of Columbus, and after Vasco de Gama doubled the Cape of Good Hope, commerce began to unfold its wings and spread them over the seas. It abandoned the backs of camels and asses, and, instead, took to ships.

From that moment England began to occupy a new commercial position. As ships began to do the carrying business of the world, the insular position of Great Britain told immensely in her favor, and in more ways than one, for, besides giving her the lead in the commercial race, which all the maritime nations of Europe at once commenced to run, it protected her oftentimes from hostile invasions to which her competitors were liable.

From having occupied the utmost verge of the world's commercial circle, England now stood in its very focus.

From the white chalky cliffs on the south coast of the island, near the Strait of Dover and the city of that name, England obtained, in early times, the name of Albion.

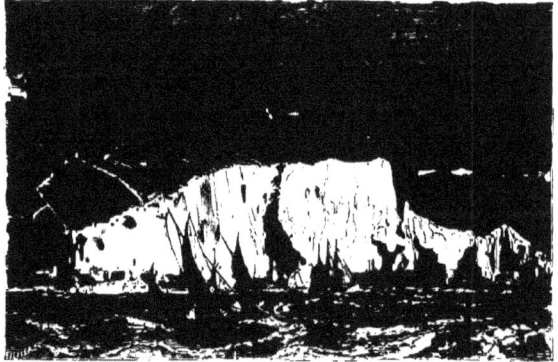

THE CASTLE AND CHALK CLIFFS OF DOVER.

4. Physical Geography.—When you study Physical Geography (and there is no branch of knowledge more interesting and instructive) you will understand how that, the very moment Columbus reported the existence of "the New World," the winds and currents of the sea conspiring, England at once became the outpost and half-way house for Europe.

You will understand how that, in consequence of the great discovery of Vasco de Gama, vessels trading to India from England could oftentimes pass the Cape of Good Hope even before their competitors from Venice, Genoa, and other Mediterranean ports could clear the Strait of Gibraltar—and how that England thus became the entrepôt between the "Old World" and the "New."

5. Industry and Resources.—The position of England being in-

sular, her population, as commerce increased, became more and more seafaring, and thus the elements of naval strength and power were placed within
her reach.

Her hills and valleys were richly stored with iron and copper, tin, lead,
and coal, and other minerals. They became the source of a great mining
industry.

In modern times, another agent has arisen which the
geographer is bound to take knowledge of in its influences upon his science. It is destined, in a measure, to
compensate other nations for the advantages which
England derived from her position with regard to the
winds and currents of the sea. The modern steamship
is to a great extent independent of these natural agencies, and has served to diminish, relatively, the naval
and commercial superiority of England.

Questions.—1. Why do you call the kingdom of Great Britain the
UNITED Kingdom?—Describe the means by which England acquired her
importance.—Why should the geography of the British Isles be so interesting
to us? *2.* How was the commerce of the world carried on before the discovery of America and of the passage around the Cape of Good Hope?—Why
was England geographically excluded from that commerce? *3.* Enumerate
the advantages which England derived from her insular position.
4. What may we learn upon this subject from physical geography?
5. In what industry did the people engage?—What is said of the English
mineral resources?—What is said of the modern steamship?

LESSON LVI.

The United Kingdom—Continued.

ENGLAND AND WALES.

The island of Great Britain is in the region of west
winds. Its shores are bathed by the warm waters of the
Gulf Stream and consequently its win-
ter climates are mild—milder than
the winter climate of South Carolina.
Look at Mercator's map and see the
difference in latitude between South
Carolina and the British Isles, and
then you will be better able to ap-
preciate the modifying influence of
this stream and of these winds upon
the climate.

In consequence of her geographical
position, the fields of England are
green all the winter through. The
country is highly pastoral as well
as agricultural. Wheat, hay, pease,
hops, and the root crops, are the chief
agricultural staples.

But, as great and important as
these branches of industry now are, and, unlimited as
her capacities now are to sustain population, yet 300
years ago, in the time of Elizabeth, she could not, without the frequent visitation of the most severe famine,
sustain a population of 4,000,000.

London, the capital of the United Kingdom, is situated on both banks of the Thames, and a few miles below the head of tide-water navigation, above which the
river becomes an insignificant stream.

Parliament sits, and the Queen holds her Court, in London. It is the
largest city in the world; it covers an area (122 sq. ms.) twice as large as the
District of Columbia, and contained, April 1, 1871, 3,880,000 inhabitants.

In England the railroads have double tracks. They are not allowed to
cross each other on the same level, as they do with us, but they are compelled to cross by going under or over. In London the cars on some lines
run over the house-tops, and on others under ground and below the cellars
and basements of the houses.

There are 28 towns in England and in Wales, 4 in Scotland, and 3 in Ireland, with a population of more than 50,000 each. Among these are Liverpool, Manchester, Birmingham, Leeds, Sheffield, Bristol, Newcastle, Stoke-
upon-Trent, Hull; in Scotland, Glasgow, Edinburgh, Dundee, and Aberdeen;
and, in Ireland, Dublin, Belfast, and Cork.

Liverpool is the grand cotton market of the world.
Manchester, only 37 miles distant, is the chief place for
its manufacture. Next to London, Liverpool is more
extensively engaged in commerce than any other city.

Birmingham is in the "Black Country"—made so by
its number of coal and iron mines. It is the great iron
mart of England

Leeds is widely known for its woollens, Nottingham
for its laces and stockings, and Newcastle-upon-Tyne
for its coal trade, glass bottles, and chemicals.

Sheffield is celebrated for its cutlery and shot-proof
iron plates for men-of-war.

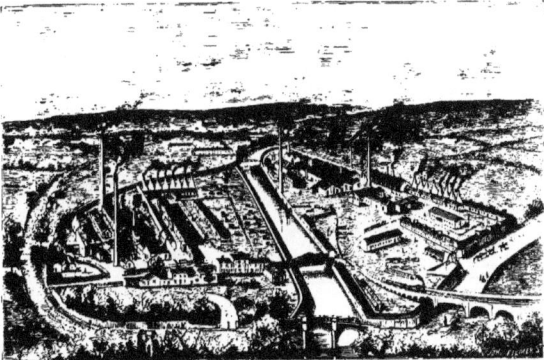

STEEL-WORKS AT SHEFFIELD.

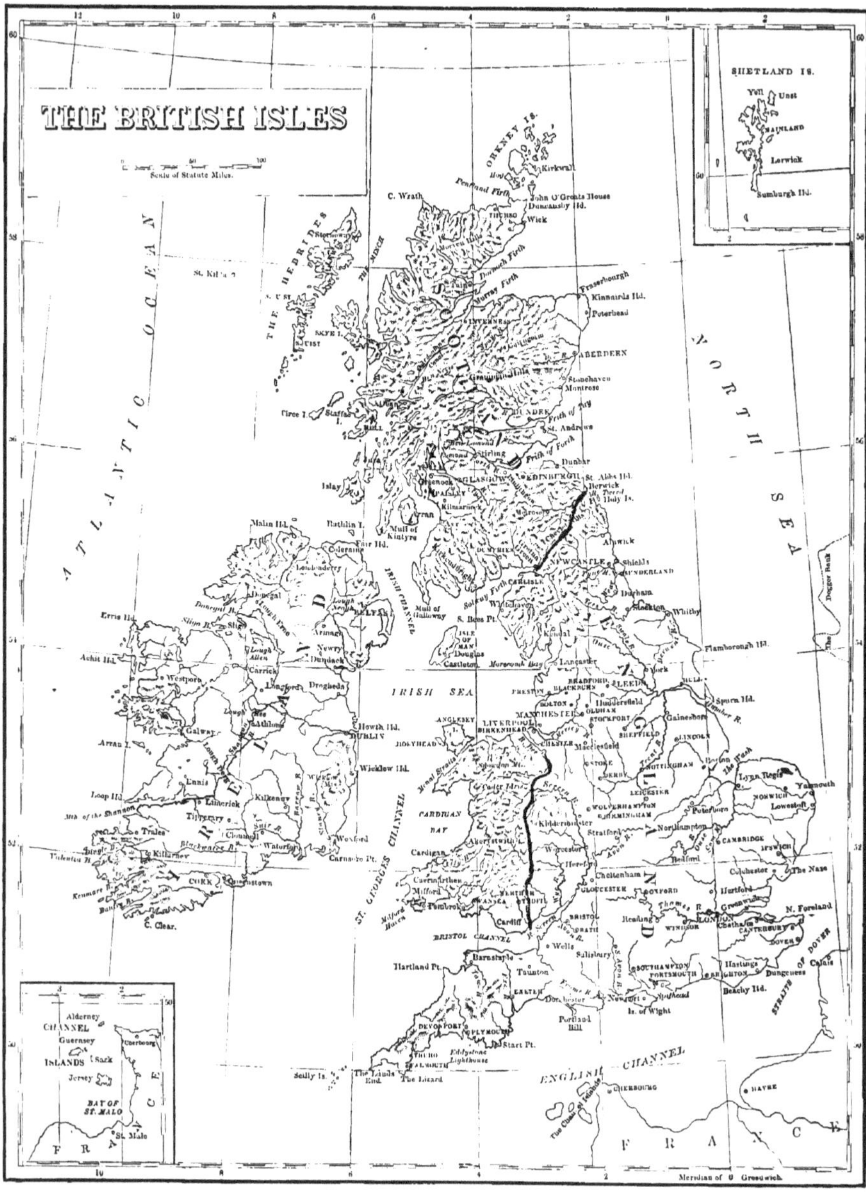

THE BRITISH ISLES

Scale of Statute Miles.

ATLANTIC OCEAN

NORTH SEA

THE HEBRIDES

SHETLAND IS.
Yell Unst
Fo
MAINLAND
Lerwick
Sumburgh Hd.

ORKNEY IS.
Kirkwall
Pentland Firth
John O'Groats House
Duncansby Hd.
THURSO Wick
C. Wrath

St. Kilda ?

SKYE I.
S. UIST
Tiree I. Staffa I.
Islay

Moray Firth
Kinnairds Hd.
Peterhead
INVERNESS
ABERDEEN
Stonehaven
Montrose
Grampian Hills
DUNDEE
Firth of Tay
St. Andrews
Stirling
Firth of Forth
Dunbar
EDINBURGH St. Abbs Hd.
GLASGOW Berwick
PAISLEY R. Tweed
Kilmarnock Holy Is.
Alnwick
NEWCASTLE Shields
NORTHUMBERLAND
Durham
Whitby

Dornoch Firth

Malin Hd.
Rathlin I.
Mull of Kintyre
Fair Hd.
Coleranie
Londonderry

Mull of Galloway
S. Ness Pt.
ISLE OF MAN
Douglas
Castleton
Morecambe Bay
Lancaster

Flamborough Hd.

Donegal B.
Sligo B. Lough Erne
Slieve R.
Achill Hd.
Westport
Carrick
LEEDS
BRADFORD
PRESTON BLACKBURN
BOLTON Huddersfield
MANCHESTER OLDHAM
LIVERPOOL Stockport
BIRKENHEAD SHEFFIELD
CHESTER Macclesfield
LINCOLN
NOTTINGHAM
DERBY
LEICESTER
Peterboro
CAMBRIDGE
IPSWICH
Colchester
The Naze

Lough Neagh
BELFAST
Newry
Dundalk
Drogheda
Howth Hd.
DUBLIN
HOLYHEAD
Wicklow Hd.
ANGLESEY
Menai Strait
Conway

IRISH SEA
ST. GEORGE'S CHANNEL

Galway
Galway B.
Lough Corrib
Athlone
Longford

Ennis
Loop Hd.
Tralee
Tipperary
Kilkenny
Waterford
Wexford
Carnsore Pt.

CARDIGAN BAY
Cardigan
Aberystwith
SHREWSBURY
WOLVERHAMPTON
BIRMINGHAM
Kidderminster
Worcester
Stratford
NORTHAMPTON
BEDFORD
Hereford
Cheltenham
GLOUCESTER
OXFORD
Reading
WINDSOR
LONDON
CANTERBURY
DOVER
N. Foreland

Carmarthen
SWANSEA
Milford Haven
Pembroke
Cardiff
BRISTOL
BATH
Wells
Salisbury
SOUTHAMPTON
PORTSMOUTH
BRIGHTON
Hastings
Dungeness
Beachy Hd.

BRISTOL CHANNEL
Hartland Pt.
Barnstaple
Taunton
EXETER
Dorchester
Newport
Is. of Wight
Portland Bill

STRAITS OF DOVER
Calais

DEVONPORT
PLYMOUTH
Start Pt.
TRURO
Eddystone Lighthouse
The Land's End
ST. AUSTELL
The Lizard
Scilly Is.

ENGLISH CHANNEL
CHERBOURG
HAVRE

FRANCE

ALDERNEY
CHANNEL
Guernsey
Cherbourg
Sark
ISLANDS
Jersey
BAY OF ST. MALO
St. Malo

Meridian of 0 Greenwich.

Portsmouth and Chatham are among the chief naval stations.

The Universities of Oxford and Cambridge are the most famous in England.

WALES is a hilly country, and its high state of improvement makes some of its landscapes most lovely. Wales is celebrated for its mines, especially those of coal and copper, a tunnel in one of which has been carried out some distance under the sea, where it is still worked.

Merthyr-Tydfil, with a population of 97,000, is the largest town in Wales.

SCOTLAND

Is the most mountainous part of the island. Its geographical position makes it also the most dreary and bleak. In winter the winds are cold, and the nights long. In the extreme northern parts there is at the time of the summer solstice no night. The sun sets, but a twilight, bright enough to read by, lasts until sunrise again.

The chief agricultural staples in Scotland are oats, barley, and the root crops.

The hills afford fine sheep-walks, and good pasturage for cattle.

Among the hills of Scotland are found those beautiful lakes which history, song, and story have made so famous. Many tourists visit the Highlands of Scotland annually, merely to enjoy the beautiful scenery and to indulge their fancy in associations around which history and romance have thrown their enchantments.

The Scotch are a steady and thrifty people, fond of learning, and given to hospitality.

In religion they are inclined, for the most part, to the Presbyterian form of worship, as the Irish are to the Roman Catholic, and the English to the Episcopal form.

Glasgow is the largest city in Scotland, but Edinburgh, on account of its traditions, its institutions of learning, and the eminent men that it has sent forth into the world, is the most illustrious.

Scotland, with a population of 3,359,000, has four large cities, viz., Glasgow, Edinburgh, Dundee, and Aberdeen.

Glasgow is the chief manufacturing town. Its industry is directed mainly to the manufacture of cotton goods, and to the building of ships and marine engines. The largest chemical works in the world are here.

Dundee is extensively engaged in the manufacture of linens, and Paisley in the manufacture of shawls.

The famous Caledonian Canal is in Scotland, and Fingal's Cave is in the island of Staffa.

The inhabitants of Scotland call themselves Lowlanders or Highlanders, according to the part of the country in which they live. Those who live near the Border resemble the English in manners, and are the Lowlanders. Those who live beyond the Grampian Hills are called the Highlanders.

NOTE.—For questions on this lesson, see end of Lesson LVIII.

LESSON LVII.

Studies on the Map of the British Isles.

Bound England.—Scotland.—Wales.—What is the shortest distance from England to Ireland?—From Scotland to Ireland?—What sheet of water separates England from France?—How wide is it?—What strait between England and Ireland?—How wide is it?

Point out and describe the position—with regard to the coast—of London, Liverpool, Portsmouth, Newcastle-upon-Tyne, Edinburgh, Dundee, Aberdeen, Glasgow, Paisley, Belfast, Dublin, Queenstown.—Which of these cities are upon rivers?

Describe the course of the rivers.—In what part of the island, and how far from London, is Birmingham?—Leeds?—Sheffield?—Oxford?—Cambridge?—Where is Manchester?—How long, from north to south, is the island of Great Britain?—Name and describe all the rivers of the island.

Mention its lakes.—Its mountains.—Channels.—Straits.— Seas.—Friths.—Bays.—Capes and headlands.—What is its greatest breadth from east to west?—What islands lie along the west coast?—What, along the north and south?—On which coast are the most islands?—Where is the Caledonian Canal?—Where is the island of Staffa?—What is it noted for?—Name and describe the rivers, capes, lakes, bays, and islands along the coast of Ireland.

On which coast are the islands most numerous?—Where is Rathlin island?—Where is Holyhead? (The Irish mail is carried between Dublin and Holyhead by the fastest sea-steamers known.)—Where are the Grampian Hills?—Name all the seaport towns on the south coast of England, and tell which way each one is from Portsmouth.—Name all the seaport towns between Beachy Head and Flamborough Head, and tell which way each one is from Yarmouth.—Name the seaport towns in England north of Whitby, and tell which way each is from Sunderland.—Name all on the east coast of Scotland, and tell which way each is from Murray Frith.

Where is John O'Groat's House?—Which way, and how far is it from Land's End?—From the Orkney Islands?—How far, and which way is it from Carlisle to Newcastle?—From Glasgow to Edinburgh?—From Dumfries to Liverpool?—Name the seaport towns of Wales.—Tell their direction from Milford Haven.—Which way from Cape Clear to the Scilly Islands?—From the Isle of Arran to Valentia Bay? (The Atlantic Telegraph extends from Valentia Bay to St. John's, Newfoundland.)

ATLANTIC CABLE—ACTUAL SIZE.

LESSON LVIII.

The United Kingdom—Continued

IRELAND.

Ireland, unlike Scotland, is for the most part flat and boggy. It is the first to catch the west winds as they come loaded from the sea with the warmth and moisture of the Gulf Stream. It is said to rain in Ireland three-fourths of the year, and the climate is very damp and mild; so much so, even in winter, that its green fields have won for it the name of the "Emerald Isle"

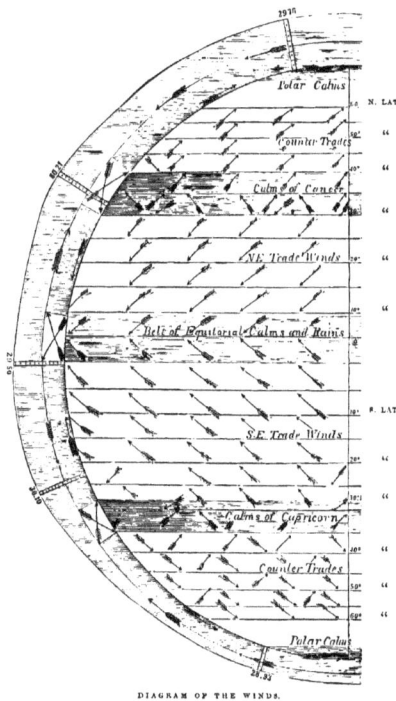

DIAGRAM OF THE WINDS.

In the above diagram the arrows fly with the wind. They show you the prevailing winds in different parts of the globe. By ascertaining the latitude of a country and locating it on the diagram, you may form an idea of its winds and its moisture, and hence of its climate.

Between the tropical circles all the winds blow from the eastward. They are called the *Northeast Trade-winds*, on the north side of the Equator; and the *Southeast Trade-winds*, on the south side. The region or belt of Southeast

Trades is broader than the region of Northeast Trades. *Examine the Diagram.*

There is near the Equator and each of the tropical circles a belt more or less marked by a calm and tranquil atmosphere. The Equatorial Calm Belt is a place of incessant rains.

The arrows crossing each other in the light spaces represent the air from the Equator and the air from the poles exchanging places as upper and lower currents, as travellers find on the Peak of Teneriffe.

Beyond either tropic, in the temperate zones, the prevailing winds are no longer *easterly* but *westerly winds.* They are the *Counter-trades*, and in the Southern Hemisphere are called "the *Brave West Winds*," they are so strong and steady. Maury's Sailing Directions taught navigators how to take advantage of them, and by so doing, the voyage for sailing-vessels bound from England to Australia and back, has been shortened several months. (See the dominant winds in Chili, p. 99.)

The numbers on the semicircumference of the diagram give the heights of the barometer at the corresponding latitudes of the globe. This diagram, made from more than a million observations, if duly studied, will prove a golden key to many mysteries of geography. (Refer to p. 69, Section 4.)

Ireland lacks the mineral resources of Great Britain. Coal is wanting, and peat is often used for fuel. The chief article of food among the laboring classes is the potato.

The "Ever Green Isle" is a fine stock country. The cattle upon its hills find abundant pasturage in winter, and the most important articles of cultivation are oats and potatoes.

The population of Ireland is 5,400,000.

Even to the present generation in Ireland famine is not unknown, and there is, among those who lack bread, a large emigration annually to the United States.

Dublin is the capital, and a fine, flourishing city, with a large commerce.

Belfast manufactures more linen goods than any other town in the British empire.

Queenstown, in the harbor of Cork, is the place where the mail-steamers that ply between Great Britain and the United States touch, to receive and land passengers and mails, both coming and going.

Questions.—Describe the situation of the British Islands, and the influence exercised by the Gulf Stream upon their climate.—Which has oftentimes the coldest weather in winter, South Carolina or Great Britain?—How do you account for it?—What is the difference of latitude?—What are the chief industrial pursuits of the people?

Describe London, and the railways of England.—How many cities are there in England having more than 50,000 inhabitants each?—Where is the great cotton-market of the world, and where the most extensive manufacture of cotton goods?—What are its chief iron and coal markets?—For what branches of industry are Liverpool, Manchester, Birmingham, Sheffield, Leeds, and Nottingham chiefly noted?—Where are the principal naval stations?—Describe Wales; its minerals and industries.—What is the largest town in Wales?—Where is it?

Describe Scotland and its inhabitants.—Their religion.—What are the chief staples of cultivation in Scotland?—What, besides their natural beauty, attracts tourists to the Highlands of Scotland?—Describe the chief towns of Scotland; their industries and population.—What is the population of Scotland?

Describe Ireland.—Explain the Diagram of the Winds.—The chief agricultural productions of Ireland.—Its population and chief towns.—Their industries, and the resources of the country.

LESSON LIX.

France.

1. Climates.—The climates of France and Western Europe depend not so much upon distance from the Equator as upon distance from the warm waters of the Gulf Stream. Thus the Gulf of Finland is closed with ice annually, from late in the fall till early in the summer; whereas, ice never forms at all in the harbor of Hammerfest, which lies 12° of latitude farther to the north.

Bearing in mind this fact, and connecting it with what you have learned about the climate and productions as affected in England by the Gulf Stream, and in America by latitude, you have at once the geographical key to the agricultural staples of almost all parts of Europe.

France has a larger population than Great Britain, but not so large as the United States. It is in area larger than California, but not so large as Texas.

It lies between the parallels of 43° and 50°. Its climate (except in so far as it is modified by the west winds which come fresh from the Atlantic Ocean, and by the south winds, which come from the Mediterranean) answers, in a measure, to the climate of Michigan, Western New York, and the Province of Ontario.

Owing to its situation with regard to these two sheets of water, the winter climate of France is not so cold as that of American countries in the same latitude; but the length of day and night, the inclination of the sun's rays, and the intensity of summer heat, are the same.

2. Productions.—The winter climate of France being milder than the winter climate of Michigan or New York, many out-door plants which cannot endure the American winter will stand the cold in France very well. Hence, France will produce everything that grows in Michigan, and more besides. It is a good grain and grass country.

In the south of France, along the shores of the Mediterranean, and under the climatic influences of that sea, the vine, the olive, the orange, the pomegranate, the fig, and the silkworm all thrive.

In the middle and north of France the beet-root, for sugar, is an important article of cultivation.

The breeding of fish and the raising of fowls also receive much care and attention. Those who are engaged in these little branches of industry earn, from the productions of eggs, fowls, and fish, a revenue of fifty millions annually.

3. Religion.—The people of France are allowed religious freedom; but the majority profess the Roman Catholic faith.

4. Cities.—Paris is the finest capital and the most splendid city in the world. It gives the fashion in dress for Europe and America, and the French tongue is the language in which the potentates of Europe converse with each other.

As designers of patterns, and in the arrangement of colors for dress, the milliners of Paris and the artisans of France excel all others.

The workshops of Paris will turn out in a mechanical and workmanlike style, anything from the finest cambric needle to the stoutest locomotive or most powerful steam-engine.

Paris excels in the manufacture of jewelry, mock and real, of gloves, perfumery, and fancy articles of all sorts. All those articles in our stores known as French goods come from Paris. This city is famous also for its schools and academies, its scientific societies, institutions of learning and public places.

Lyons is the second city in France, and the foremost of all in silk manufactures. It employs 100,000 persons in this industry alone.

Marseilles is the largest seaport town in France.

It is extensively engaged in the manufacture of various articles, from soap to steam-engines. It also has a large trade in human hair.

Marseilles is on the Gulf of Lions, or Gulf of the Lion, so named, not from the city of Lyons, but from the exceeding storminess of its waters.

Bordeaux is noted for its wines; Lille for its manufacture

GRAPE-GATHERING IN FRANCE

of flax and cotton goods, beet-root sugar, and of rape-seed and linseed-oils.

Toulouse is famed for its steel-works, cannon foundries, and its woollen-factories, and Rouen for its cotton-mills. It is the Manchester of France. St. Etienne is in the midst of coal-mines. It manufactures about twelve millions of dollars' worth of ribbons annually. It is also extensively engaged in making fire-arms, bayonets, and cutlery.

Toulon is the great naval dockyard of France ; about 6,000 hands are employed continually in it.

The winters at Toulon are so mild that the fig, date, orange, aloe, and pomegranate flourish in the open air.

Strasbourg derives its importance from its geographical position in a military point of view. The cathedral in Strasbourg has the tallest spire in Europe, and is celebrated for its colossal clock, which, as it tells the hours, amuses the people with its puppet-shows. (Now included in Germany.)

Amiens is noted for its cotton and woollen goods.

Nismes is commercial and manufacturing, producing silks and woollens and exporting them to foreign markets, together with the oil and wine of Languedoc.

Rheims is noted as the scene of the baptism of King Clovis, in 496. Here all the sovereigns of France were crowned. It has a famous old cathedral. It is now largely engaged in the manufacture of woollens and in the sale of wines.

Cherbourg, with its fine breakwater, is the most important naval station of France, on the Atlantic.

CHERBOURG AND ITS BREAKWATER.

Montpelier is celebrated for its chemical works.

In Bayonne the first bayonets were used, hence their name.

Nantes (nantz'), where the famous edict in favor of the Protestants was signed by Henry IV., in 1598. is largely engaged in commerce and ship-building, also in sugar-refining and the manufacture of glass, cotton goods, and machinery of various kinds.

Questions.—1. Describe the climates of Western Europe, and tell the influences which modify them.—What is said of the population of France ?—Its area ?—Its climate ? *2.* Productions ? *3.* Religion of France ? *4.* Describe the chief city and capital of France—Industries in Paris.—What is said of the French language?—Describe Lyons—Bordeaux—Toulouse—Cherbourg—Rheims, and other towns.

Note, to the Teacher and Scholar.—Hereafter, no questions except those on the map will be appended to the lessons. It is believed that, after the scholar has advanced as far as this point, the questions are unnecessary and may become tedious. *Review Questions,* however, will be found at the end of the lessons on Europe, Asia, and Africa.

LESSON LX.

Austria, Hungary, and European Turkey.

1. *General Geography.*—These countries lie between the same parallels with France. They are, however, farther removed from the sea. Their climates are continental, like ours, and resemble the climates in corresponding latitudes of the United States and Canada more nearly than they resemble those of France. These countries lie between the parallels of 40° and 50°, and are identical in latitude with the States of Pennsylvania, New York, Michigan, and parts of the province of Ontario.

These countries embrace together an area of 443,000 square miles, and a population of 55,000,000 inhabitants

Excepting their capitals, they have only ten cities which contain a population of over 50,000.

This part of the continent, with the Black Sea basin, is the granary of Europe. It produces corn, wheat, tobacco, and everything else that is produced between like parallels in the United States.

AUSTRO-HUNGARIAN EMPIRE.

2. The inhabitants of this empire are Sclaves, Germans, and Magyars (*mad'jars*).

The Germans, though in the minority, are the dominant race. Theirs is the language of the empire. The emperor of Austria is king of Hungary.

The Germans are remarkable for their blue eyes, yellow hair, and fair skins. In old times they were often spoken of as "the yellow-haired Germans."

Austria is richer in minerals than any other continental nation of Europe.

MINING SALT.

Wielicza, in Galicia, is famous for its salt-mines, which have been worked for more than 600 years. There is a chapel in the mine dedicated to St. Anthony. It was built more than 400 years ago. Its walls, ceilings, and floors; its columns, with their ornamental capitals; its altars, and images, and chancel, are all carved out of the natural salt. Lemburg, the capital of Galicia, is noted for its fairs.

The quicksilver-mines of Idria, in the neighborhood of Laybach, are second only to those of Almaden in Spain, the most celebrated in the world.

The hills of Austria, which abound in almost every kind of metallic formation, are rich in mineral waters also. Those of Carlsbad, Toplitz, and Seidlitz are of world-wide celebrity.

During the great Earthquake at Lisbon in 1775, the waters of Toplitz and Carlsbad became turbid, then ceased temporarily; afterward they gushed out in blood-red color.-It is said.

Bohemia is also famed for its mines, its waters, and its fruits. Forests of damson and plum-trees are found there.

You have, no doubt, heard of the Bohemian glass.

The Jewish population of Austria, Hungary, and Poland is very large. They, for the most part, carry on the trade of the country.

15

Vienna, on the Danube, with a population of 607,000, is the capital, and the chief seat of the manufacturing industry of the Austrian empire. It is the principal focus of its inland trade, as Trieste is of its foreign trade.

The workshops and artisans of Vienna, send forth annually large quantities of hardware, porcelain, silks, jewelry, gold and silver embroideries, and musical instruments elaborated with much taste and skill.

Prague, with a population of 142,000, has the oldest University in Germany. It was the birthplace of John Huss, Kepler, and many other eminent men. It is extensively engaged in manufactures, which are of the coarser sorts, and are intended chiefly for the inland markets of Germany and the neighboring States.

Gratz, the capital of Styria, with 63,000 inhabitants, is in a rich country, and affords the cheapest living of any place in Europe.

Brünn, with 58,000 inhabitants, is to Austria what Leeds is to England for woollens, and Lyons is to France for silks, Belfast to Ireland for linens, and Pittsburg to Pennsylvania for glass.

Pesth, with a population of 131,000, on one side of the Danube, and Buda (sometimes called Ofen, the oven, from its hot springs) on the other, with a population of 55,000, form Buda-Pesth, the capital of Hungary. The celebrated crown of St. Stephen, which was given to him by the Pope in the year 1000, is kept here.

HUNGARIAN COSTUMES.

The nobility sometimes own, as in feudal times, immense estates. From the banks of the Danube all the way up in the direction of Brünn toward the head-waters of the Vistula and the Oder, the country, almost without interruption, for the distance of 200 miles, is the private property of a single Austrian nobleman.

TURKEY IN EUROPE.

3. The Turkish empire lies, a part in Africa, a part in Asia, and a part in Europe.

	AREA.	POPULATION.
Turkey in Europe,	110,689	9,213,762
Turkey in Asia,	667,244	17,000,000
Turkey in Africa,	952,830	9,000,000
Total,	1,730,763	35,213,762

Turkey in Europe is semi-peninsular. It has a sea on the east and on the west, and, at no great distance, on the south. This modifies the climate. See p. 62.

The Turks, like the Arabs, have dark complexions.

Constantinople, the great Turkish emporium, was founded by Constantine the Great in 328, on the site of old Byzantium. He was the first Roman emperor professing Christianity. Eleven centuries and a quarter afterward it fell into the hands of the Turks. They have held this city ever since. They have converted the splendid Christian church of St. Sophia into a magnificent mosque. The seraglio (se-ral'yo) of the Grand Turk adjoins it.

The geographical position of Constantinople confers such advantages upon it, in a military point of view, as to make it, in the eyes of the great powers of Europe, "the Key to the East."

Generally, the streets of this city are narrow and filthy, and the private houses are mean, and built chiefly of wood; hence the frequency and destructiveness of fires in Constantinople. It is said to be burnt down on the average once in every fifteen years. The city is given up to idleness and luxury.

It is computed that there are no less than 80,000 wherries plying daily for

THE BOSPORUS.

hire on "the Golden Horn" and its adjacent waters. "The Golden Horn" is an inlet of the Bosporus. Its chief manufacturing industry is directed to meerschaum pipes and leather. (The word meerschaum means the foam of the sea. It is really a kind of soft chalk.)

The population of Constantinople is 1,075,000.

Adrianople is in the land of roses; plantations of them are cultivated for their essence. Like Constantinople, and most Turkish cities, Adrianople is beautiful in the distance, but when the traveller enters it, the enchantment which distance lends is gone.

Bucharest, the capital of Roumania, exports vast quantities of grain, wool, timber, salt, wax, and other raw produce.

Salonika, with 90,000 inhabitants, is also beautiful in the distance, with its mosques, minarets, domes, and towers. It, too, has a large trade based on the exportation of raw produce. Bosna Serai has 122 mosques.

The celebrated Mount Athos stands on a peninsula to the east of Salonica. It has been occupied from time immemorial by a community of Greek monks, who pay the Sultan an annual rental of about $20,000. They are governed by one of their own order styled "the first man of Athos," and one of the rules of this order is, that no woman or any other female creature, not even a cow or a hen, shall enter their domains. Their villages are inhabited entirely by bachelors.

The government of Turkey is styled the Sublime Porte, and its sovereign the Sultan or Grand Seignior. Many of his subjects are Jews.

The Danubian provinces of Servia and Roumania, and the mountain principality of Montenegro, by the treaty of Berlin (1878), were made independent of Turkey. The new provinces of Bulgaria and East Roumelia have Christian Governors, but are partially dependent on Turkish rule. In Servia there is no distinction of classes. Nearly every family has a freehold. There are no paupers. They have trial by jury; no established church, but toleration for all. The country abounds in natural resources.

LESSON LXI.

Greece and Italy, Spain and Portugal.

1. General Geography.—These three peninsulas lie between the same parallels of latitude. They have corresponding climates, and, consequently, similar industrial pursuits. They have an aggregate population of 46,480,000.

There is no city in Greece with a population as great as 50,000.

2. Cities of the Spanish Peninsula.—In the peninsula of Spain and Portugal there are, besides the capital of Spain, only four inland towns with more than 50,000 inhabitants, and two Granada and Saragossa owe their present, though declining, proportions, rather to their ancient renown, their traditions, and the prestige of their former glory, than to present industries or any living spirit of enterprise among their inhabitants.

Each was once the capital of a powerful empire. Saragossa—a corruption of its Roman name, "Cæsar Augustus"—was in the days of Rome a noble city.

In 1808–9, when the French made an aggressive war upon Spain, this city was made illustrious by the noble stand which the inhabitants made for its defence.

It was the capital of Aragon, of which the patroness of Columbus—Isabella—was the queen, at the time of the discovery of America.

The glory of Granada has also departed. The splendid alabaster monuments of Ferdinand and Isabella are there. It was the last stronghold of the Moors in Spain.

3. The Alhambra was the great Moorish palace, and it is the finest specimen of arabesque architecture in Europe.

4. Gibraltar is a celebrated fortress cut out of the solid rock. It commands the passage between the Mediterranean and the Atlantic. The town is inhabited by people of all nationalities. It is a free port, and belongs to England.

5. Alpine Passes.—One of the best passes across the Alps leads through Milan, the entrepôt of trade from Genoa and Northern Italy with Central Europe.

There is a considerable Swiss trade with Italy across Lake Lucerne and over the pass of the St. Gothard. The Great St. Bernard (crossed by Napoleon and his army in 1800), the Grimsel, the Cervin, and the Splugen are the other passes most frequented.

Turin, like Milan and most of the cities in this part of the country, live upon their ancient renown. Its manufacturing industry is chiefly in silk.

6. The Mont Cenis Tunnel.—By means of the Mont Cenis Tunnel, which is cut through the Alps, and is the grandest work of the kind in the world, Turin is connected with all parts of France and Germany.

MONT CENIS TUNNEL.

7. Staples.—The staple productions of these three peninsulas are the same. Corn, wheat, the olive, and the vine, the silk-worm and fruit, all thrive equally well in them.

8. Spain and Portugal are both mountainous countries, many of the peaks being high enough to be always covered with snow. In the last century they were both ranked among the first-class powers of Europe. They failed to catch the spirit of progress and improvement, however, which mark the age, and have consequently dwindled down into second and third rate powers.

Both of these countries profess the Roman Catholic religion. Their inhabitants are of a dark complexion, with black hair and eyes, quite different from the yellow-haired, blue-eyed Germans. But like the Italians, and all other people who live in mild climates and under bright skies, they are lively and gay; prone to out-door amusements, fond of bright colors, and much given to music and dancing.

The chief articles of export are wine and oil, olive, aloes, and other fruits, both fresh and dried. The Andorra Republic has 150 sq. ms.; pop. 12,000.

Our merchants have quite a number of vessels employed especially to bring raisins, grapes, and oranges from Malaga, and there is a large trade also with England in those fruits.

Madrid, with a population of 476,000, is the capital and largest city of Spain. Barcelona, with 252,000 inhabitants, is next in size.

Spain is rich in minerals, and among the most famous mines in the world are the quicksilver mines of Almaden—famous for their yield, and the centuries for which they have been worked.

Lisbon is the capital of Portugal, with 225,000 inhabitants. It was the scene of a fearful earthquake in 1755.

ITALY.

9. Italy was the seat of ancient Rome. It was the land of Galileo, Dante, and Tasso.

From her peninsular position, Italy, as did England in her insular position, stood comparatively secure from outside attacks in olden times.

Italy now contains Piedmont, Lombardy, Tuscany, Naples, Sicily, Sardinia, and the States of the Church.

The ruins, the traditions, and the associations connected with Rome make it famous, and continually crowd it with students, scholars, learned men, and tourists.

Rome is on the classic Tiber. It has been lately seized by the Italians and occupied as their capital. This renowned city is in latitude 41° 50', and has now a population of 220,000.

NOTE.—Let the pupil compare the latitude of Rome with that of some of the capitals of the great nations which have made their mark in history, and note their difference in climate.

Italy is also Roman Catholic in religion. The Pope still holds his court in Rome; though his temporal power is limited to the States of the Church, with a population of 724,000, and an area of 4,550 square miles. The Pontine Marshes are in the southern part of the pontifical territory. They were once so poisonous, that to sleep in them for a single night was considered fatal.

Venice, with 120,000 inhabitants, occupies 72 small islands, connected by bridges. The streets are navigated in gondolas.

Naples, with 420,000 inhabitants, is the largest city of Italy.

Italy grows the mulberry, the olive, and the vine; rice, cotton, and the cereals are also cultivated there.

Lombardy is the best-cultivated part of the country. Its system of irrigation is perhaps unsurpassed. It is also a fine cattle country and famed for the produce of its dairies.

Iron and sulphur, boracic acid, and marble are the chief mineral productions of Italy.

The climate of Italy is mild, and oranges and lemons ripen, in some parts, as early as March.

From the marble-quarries of Greece and Italy the artists of all ages have cut their finest statues. The picture-galleries of Italy are very famous too, and they attract young artists from all parts of the world.

The volcanoes of Etna, Stromboli, and Vesuvius are in the domains of Italy.

The cities of Herculaneum and Pompeii were swallowed up in the lava and ashes from Vesuvius hundreds of years ago.

The government of Italy is causing these cities to be excavated, and the workmen find the bodies of the bakers at their ovens, of the potter at his wheel, of the servant in the kitchen, of the mistress in the parlor, and all the inhabitants at their various occupations, just as they were at the moment of the great catastrophe.

San Marino is a little republic set on the top of a rugged mountain over 2,000 feet high.

Genoa, famous as the birthplace of Columbus, manufactures the finest of silks and velvets.

GENOA.

GREECE.

10.—Greece, like Rome, was also once the seat of an empire that ruled the world.

The peninsula of Greece, with its islands, is about half as large as South Carolina. It contains 1,097,000 inhabitants, many of whom are brigands.

Greece suffered under Turkish misrule for many generations. At last, in its struggle for independence forty years ago, it was assisted by the great powers, and erected into a separate kingdom under their guarantee.

It has a large seafaring population. Greek merchants and Greek sailors are to be found in all the ports of the Mediterranean. Its industries are chiefly agricultural.

We get from Greece, currants, figs, and other dried fruits. Attention there is paid to the cultivation of the vine and the mulberry, and to the making of wine and the manufacture of silk goods. But the country is infested with highway robbers and brigands, and the laws are badly administered.

Athens, with a population of 40,000, is the capital of Greece, and is unrivalled for the fame of its ancient philosophers, poets, painters, and orators.

The Ionian Republic comprises seven islands.

Corinth was in ancient times famed for its wealth, its splendid edifices, and its moral corruption.

LESSON LXII.

The New German Empire and the Smaller Powers.

1. General Geography.—The smaller Powers are the Kingdoms of Holland, Belgium, and Denmark. These three Kingdoms form no part of Germany. The new *German Empire* consists of the Kingdoms of Prussia, Bavaria, Würtemberg, and Saxony, 13 Duchies, 6 Principalities, 4 Free Cities, an Electorate and a Landgraviate, with the newly-acquired provinces of Alsace and Lorraine; but all these States, with Belgium, Holland, and Denmark, embrace an area of 265,000 square miles, with a population of 51,000,000.

It is worthy of note, that these 51,000,000 people all inhabit a region of country not quite so large as Texas.

The total population of the New German Empire is 41,000,000; its area is 225,000 square miles.

These 51,000,000 people have built up 31 cities, each with a population of 50,000 and upward.

There is a large emigration annually from these countries to the United States, chiefly from the ports of Bremen and Hamburg, from which there are lines of steamers to America, as there are from Havre, Southampton, and Liverpool.

2. Hamburg and Bremen.—Hamburg, with a population of

305,000, is the chief seaport of the North German states. It is on the Elbe, about 70 miles from its mouth.

Bremen has a population of 75,000, and is a rival of Hamburg for German commerce.

Lubeck, another of the Free Cities, next in importance to Bremen, has a population of 51,000. The crust of the earth at Lubeck has been, in the course of a few centuries, perceptibly raised out of the sea.

3. Intellectual Character.—The inhabitants of these countries are highly intelligent, and very industrious, ingenious, and thrifty.

To promote trade, great commercial fairs are held at stated periods and at various places. Among the most famous are those held at Leipsic. This town is celebrated for its type-foundries, printing establishments, and the cheapness, variety, and numbers of its publications in all languages.

PRUSSIA.

4. Prussia is the master-spirit of Germany.

Berlin the capital of Prussia and also of the German Empire, has a population of 1,000,000, and is one of the finest cities of Europe.

ROYAL PALACE AT BERLIN

At Berlin resides William, the first emperor of the new German Empire.

In the war of 1866, which lasted but seven weeks, Prussia completely absorbed the Kingdom of Hanover, the Electorate of Hesse-Cassel, the Duchy of Nassau, the free city of Frankfort-on-the-Main, and the Landgraviate of Hesse-Homburg. At the same time, the Duchy of Holstein and the Duchy of Schleswig, once Danish, were annexed to Prussia; twelve States, free cities, viz., the Kingdom of Saxony, the Grand Duchies of Mecklenberg-Schwerin, Mecklenberg-Strelitz, and Oldenberg, the northern half of the Grand Duchy of Hesse-Darmstadt, the Duchies of Brunswick and Anhalt,

the principalities of Lippe and Waldeck, the six Thuringian States, and the Reuss States, all entered into most intimate and dependent relations with Prussia. Bavaria, Baden, Wurtenberg, and the southern portion of Hesse-Darmstadt, entered into a treaty with the King of Prussia, agreeing, in time of war, to put their armies under his control.

These States have fine climates and a productive soil, under which is stored away mineral treasures of great value and abundance. They are rich also in mineral springs and medicinal waters of various sorts and temperatures. The most famous watering-places are the hot springs of Baden-Baden, the warm springs of Aix-la-Chapelle, and the boiling springs of Wiesbaden.

The most famous mineral production of Prussia is the zinc of Silesia, with which the markets of India are supplied. But the most singular production is amber, which is cast up by the waves of the Baltic, and the collection of which is the prerogative of the crown.

Königsberg and Stettin are important seaports.

Cologne, with 126,000 inhabitants, is the most populous city of Prussia, on the Rhine. Ehrenbreitstein, opposite Cologne, is one of the strongest Prussian citadels. From Dantzic, large quantities of grain are shipped.

Frankfort, with a population of 80,000, was the capital of the old Germanic Confederation. It stands on the right bank of the Main. Munich, the capital of Bavaria, is renowned for its literary institutions and its galleries of art.

5. SAXONY is celebrated for its breed of sheep, and its wool.

6. BELGIUM is the best cultivated and most densely populated country in Europe.

Brussels, the capital of Belgium, on the Senne, with 190,000 inhabitants, is widely known for its lace and its carriages. Ghent nearly equals Brussels in population, but Antwerp is the great commercial emporium of Belgium, and has 125,000 inhabitants.

Leige, with 97,500 inhabitants, is the seat of the Belgian iron-works. Ostend is an important seaport on the North Sea.

7. HOLLAND has literally been reclaimed from the sea, the waters of which are kept out by means of embankments called dykes. It is a low and flat country, intersected with canals as others are with roads.

Holland is a sort of a dairy-farm, from which the markets of Hull, London, and other English towns are chiefly supplied, especially with cheese, beef, cattle, and butter.

The Dutch are a sober, provident, and industrious

people. They are fine sailors, and are extensively engaged in the sea-fisheries as well as in commerce and navigation.

Holland used to be a great naval and commercial power. She has large possessions both in the East and West Indies.

The chief branches of industry here are connected with the soil and the sea. The most celebrated lapidaries are in Amsterdam.

AMSTERDAM.

Amsterdam is the largest and most important town in Holland, with 265,000 inhabitants.

The Hague is the capital of Holland.

LESSON LXIII.

Switzerland, Norway and Sweden, and Denmark.

SWITZERLAND.

1. The Republic of Switzerland, though only one-third larger than Maryland, consists of 22 separate States or "Cantons," which are as distinct from each other as are the States in the American Union.

Their independence dates from 1307, and the affairs of the Republic are, as with us, managed by a Congress.

Regarding Europe as a watershed that is drained, as its rivers show, off to the north, south, east, and west (study the Orographic View of Central Europe), Switzer-

land, with its Alps, is at the top of the roof, for you see that the great rivers, as the Danube, Rhine, Rhone, and Po, rising in these mountains, empty into the Black Sea, the North Sea, the Mediterranean, and the Adriatic.

2. Mont Blanc, 15,740 feet high, is the highest peak in Europe, and, though situated within the borders of France, belongs to the Swiss Alps. Mont Cervin, or "*The Matterhorn*," the rival of Mont Blanc and Monte Rosa, is an imposing obelisk, and has defied every attempt to scale its ice-clad pinnacle.

These mountains are celebrated for their snow-clad peaks, beautiful lakes, grand scenery, and their glaciers, which are immense masses of ice and snow that are always sliding down from the mountains into the valleys below. As they move, rocks, soil, trees, and everything are overwhelmed and carried down before them. Some of the most celebrated philosophers of Europe have spent much time there in studying these curious phenomena.

More than 1000 square miles are covered by the glaciers, and they are estimated to vary in thickness from 200 to 5000 feet.

Switzerland is the most mountainous country in Europe.

The celebrated institution of charity, the Hospice of Mount St. Bernard, is in Switzerland. It is situated 8,185 feet above the level of the sea, and is the highest place of human habitation in Europe.

The Hospice of St. Bernard has been occupied for ages by Benedictine Monks, whose business it is to refresh and relieve travellers. They teach the celebrated dog of St. Bernard to hunt and relieve wayfarers who get lost in the snow.

These sagacious animals are always sent out during a snow-storm with baskets of provisions and wine tied around their necks to relieve and revive those who are perishing in the bitter cold of these mountains.

In the dead-house are contained the frozen and unrecognized bodies of travellers who have perished in the snow; they never thaw, and lie there like so many statues of marble.

The Swiss are said to be the Dutchmen of the mountains; they are phlegmatic, industrious, and liberty-loving.

Owing to the mountainous character of their country, which is better adapted to grazing than to tillage, they do not produce breadstuffs enough for their own use, and their industries are devoted chiefly to cattle-raising, dairy-farming, and manufacturing.

3. We get from there Swiss muslins, ribbons, toys, and carved work. They excel in the manufacture also of watches and musical-boxes. Geneva is famed for these.

OROGRAPHIC
VIEW OF
CENTRAL EUROPE

ATLANTIC OCEAN

MEDITERRANEAN SEA

BAY OF BISCAY

ENGLISH CHANNEL

NORTH SEA

BALTIC SEA

GULF OF LIONS

TYRRHENE SEA

ADRIATIC SEA

IONIAN SEA

AEGEAN SEA

BLACK SEA

"THE MATTERHORN" AND ITS GLACIER.

[List of names answering to numbers on Orographic View of Central Europe.

1. St. Petersburg.	9. Stockholm.	16. Turin.
2. Moscow.	10. Copenhagen.	17. Paris.
3. Constantinople.	11. Oldenburg.	18. Madrid.
4. Vienna.	12. Bremen.	19. London.
5. Berlin.	13. Berne.	20. Edinburgh.
6. Prague.	14. Florence.	21. Dublin.
7. Dresden.	15. Rome.	22. Athens.
8. Munich.		

NOTE.—Let the pupil carefully find all of these places and observe their comparative elevations.]

Switzerland is classed among Protestant nations.

There are 22 Cantons in Switzerland, of which Berne, with a population of 30,000, is the capital.

SWEDEN AND NORWAY

4. Are two governments under one king. Each country makes its own laws. They form a mountainous peninsula, often called SCANDINAVIA. They lie between the parallels of 55° and 72° north latitude.

This country is so rugged, and the rays of the sun strike so feebly into the deep glens, that the cultivation of the soil is not very remunerative.

In the northern part of Norway, during several weeks in summer, the sun does not set at all, and travellers often go to the North Cape there just to see the midnight sun. It is a very curious and most interesting sight.

The chief branches of industry are connected with the forests, which are very extensive and furnish timber for the navies of Europe ; with the mines, which furnish our markets with the beautiful Swedish iron ; with navigation, that furnishes the navies of the world with the best of sailors ; or with the sea-fisheries, which furnish the chief supplies of fish for Europe.

The Swedes and Norwegians are Lutherans in religion. They therefore belong to Protestant Europe. The Mormons of Utah are recruited chiefly from Norway and Wales.

Lapland, a cold region of Sweden, Norway, and Russia, is inhabited by the diminutive Lapps.

Hammerfest, within the Arctic Circle, the most northerly town of Europe, consists of a single street of detached, one-storied frame-houses An obelisk on the outskirts marks the end of the meridian line of 25° 20', measured by the geometers of Norway, Sweden, and Russia, from a point on the Danube.

It is common for the poorest Lapps to possess a dozen reindeer, while occasionally a herd of a thousand is owned by a single individual.

A LAPP LADY.

The Lapps are highly intelligent, are not lacking in literary culture, and are remarkable for their sweetness of expression.

Stockholm, with a population of 140,000, is the capital of Norway and Sweden. It is built on a few small islands near Lake Malar.

Gottenburg, and Carlscrona, the naval arsenal of Sweden, are commercial cities.

The Norwegian city of Christiana has 40,000 inhabitants.

Frederickshall, with 7,500 inhabitants, stands on a bay of the Christiana fiord, and has an obelisk to mark the spot where the famous Charles XII. fell, in 1718, while besieging the fortress.

Bergen, with about 26,000 inhabitants, is chiefly engaged in the Lofoden fishery.

DENMARK.

Denmark is one of the oldest States in Europe. It consists of a peninsula projecting towards the southern coast of Norway, and comprises Jutland and several important islands near the entrance of the Baltic Sea.

At no very remote period the peninsula was largely covered with forests, and wolves were common and in great numbers.

The coasts of Denmark are in some places low, and dykes are necessary to keep out the waters of the sea. To its peninsular form its climates owe their humidity, evenness, and mildness.

The government is a monarchy, and the religion of the people is Lutheran.

The Duchies of Schleswig Holstein and Lauenberg once belonged to Denmark, but were wrested from her in 1864, and in 1866 ceded to Prussia.

Copenhagen, the capital of Denmark, with a population of 160,000, is an elegant and important city. Its University library contains 100,000 volumes.

It was on his way to attack Copenhagen in the winter of 1658, that Charles X. of Sweden, against the warnings of his ablest generals, marched his whole army, horse, foot, and artillery, over the frozen Baltic.

Elsinore, on the Danish Sound, is an important naval station.

LESSON LXIV.

Russia. (M. p. p. '24.)

1. Russia includes all of northern Europe not heretofore treated, and the whole of northern Asia.

Russia in Europe lies chiefly between the parallels of 45° and 70°, and Russia in Asia, between those of 50° and 78°. In the two continents it embraces an area far greater than that of the United States.

Its climates are cold and inhospitable; its geographical position is such as to exclude most of its inhabitants from navigation and the pursuits of the sea.

Throughout the vast area included within the domains of this empire only five cities of 100,000 inhabitants and upward are to be found.

2. The chief industries of Russia are connected with the land rather than with the sea. They are mining, agricultural, and manufacturing. But the productions of her factories are intended chiefly for home consumption.

3. Russia in Europe is, for the most part, a level country, sloping as the rivers flow. (See Orographic View.) That part of it which lies south of 50°, and borders on the Caspian and Black Seas, is agriculturally very rich; the soil is black and warm. It forms the country known as the "Black Lands of Russia," and resembles the prairies of Minnesota and other Western States. This section of Europe is very fertile; the soil is inexhaustible, yielding annually, without manure, two crops, a green crop and a cereal. Its wines are also very fine.

4. This part of Russia, with European Turkey, and Eastern Asia, is the granary of Europe. The corn from these regions meets in the markets of Great Britain and Western Europe the grain and breadstuffs from the United States, and thus the farmers of the far West find in the markets of London—which is the greatest grain-market in the world—competition from the far East.

Russia also sends to Great Britain hemp, tallow, leather, skins, and furs in large quantities.

5. Russia has the largest population of any State in Christendom. In Europe and Asia together she has a population of 77,000,000, and an area of 7,862,585 square miles. Russia, Great Britain, and the United States, exercise dominion over one-third of the land surface of the earth, and one-fourth of its inhabitants. The English possessions lie chiefly in warm climates, the American in temperate, and the Russian in cold. Russia has but fourteen cities of more than 50,000 inhabitants; thus indicating that her people live rather by tillage and pasturage, than by manufacturing and ocean commerce.

St. Petersburg, the capital, has a population of 550,000; Moscow 360,000; Warsaw 245,000; Odessa 120,000; Riga 100,000.

6. The Steppes of Russia extend along its southern borders all the way from the foot of the Carpathian Mountains, in Europe, to the borders of China, in Asia. This is a prairie country, and but for its distance from the sea and its summer droughts, would be as smiling as ours is.

7. Russia has a considerable inland trade with the bordering States both in Europe and Asia, and it is carried on by caravans, and disposed of by means of fairs. the most famous of which are held at Nijnii (*nizh'-nii*) Novgorod. Merchants from China, Mongolia, India, Afghanistan, Persia, Turkey, and from all parts of Europe attend this fair. The annual concourse ranges between

16

two and three hundred thousand merchants with their followers.

RUSSIAN TRAVELLERS.

8. Russia is rich in minerals of all sorts, from gold, silver, platinum, and precious stones down. There is no malachite more exquisitely beautiful than that of Russia.

9. Until lately about one-third of the people in Russia were slaves, called serfs. But the present emperor, yielding to the spirit of the age, has emancipated them.

10. Moscow is celebrated for a vast structure—comprising forts, barracks, palaces, churches, and cathedrals all in a group—called the Kremlin.

It has also been made famous by Napoleon I., who, in 1812, marched his grand army into Moscow for winter-quarters. The inhabitants of the city set fire to their houses, the whole place was laid in ashes, and the invading army was compelled to make a winter retreat, and was completely destroyed.

Moscow lies four hundred and eighty-five miles south by east of St. Petersburg. Peter the Great made the journey in winter, in an open sledge, in forty-six hours; the Emperor Alexander more than once made it in forty-two hours; the express railway train now accomplishes it in twenty-two hours.

The Suez Canal gives Russia a new route to "the East." The Odessa Company has already 80 steamers in that trade.

Kazan has a population of 58,000.

Sebastopol, famous for its siege and capture by the French, assisted by the English, in the Crimean war, 1854, has a reduced population.

Warsaw, the ancient capital of Poland, is the emporium of trade for Russian Poland.

Archangel, with a population of 20,000 inhabitants, is the great shipping port of Northern Russia, as Odessa, on the Black Sea, is of Southern Russia.

Archangel is situated close to a geographical line which marks the northern limit of cereal and garden culture; all

its grain and vegetable supplies, as well as fodder for cattle, are transported from a distance.

Astrakhan, with a population of 45,000, is an island city on the Volga river. Through it passes the trade of Russia with Western Asia.

The Russians adopt the Greek form of worship, and the emperor is the head of the Church.

11. The government is an absolute monarchy. It has no constitution, no parliament, no congress.

12. The inhabitants consist of many races, but the majority are Sclavonic, though the Teutonic, the Finnish, and the Turkish races are represented among them.

Ethnography will teach you the early history, types, and races of mankind. From it, when you study Physical Geography, you will learn that the Sclavonic race includes the Russians, the Poles, and the inhabitants of Croatia, Servia, Illyria. The Teutonic family includes the Germans, English, Dutch, Flemings, Danes, Swedes, and Norwegians. French, Italians, Spaniards, and Portuguese make up the Latin race.

13. FINLAND is a dependency of the Russian Empire, and is a Grand Duchy. Its interior is a vast plateau. The Governor-General, representing the Emperor of Russia, resides at Helsingfors.

HELSINGFORS.

LESSON LXV.

Studies on the Map of Europe. (pp. 124, 125.)

Boundaries.

Between what parallels of latitude and meridians of longitude does Europe lie?—What natural boundary has it on the north?—On the east?—On the south?—On the west?—What, on the southeast?—What grand division on the east?—What, on the south?—How do the peninsulas of Europe generally project?

Bound France.—Between what parallels of latitude does it lie?—How is it separated from England?—How, from Italy?—How, from Spain?—Bound "*the Peninsula*" (Spain and Portugal).—Bound Portugal.—Bound Italy—Greece.—Bound Switzerland.—Bound Norway—Lapland—The Empire of Austria.—Between what natural and political boundaries does Russia in Europe lie?—Bound Russia in Asia.—Bound Turkey—Italy—England—Belgium—Greece.

Where is Holland?—Denmark?—Wales?—Ireland?—Iceland?—Hanover?—Saxony?—Bavaria?—Bohemia?—Hungary?—Roumania?—Roumelia?—Servia?

Mountains.

What great range of mountains traverses a large part of Europe?—What mountains on the east?—What mountains northwest of the Black Sea?—What mountains lie northeast of the Black Sea?—Point out the Balkan Mountains.—Where are the Apennines?—The Pyrenees?—What mountains are in "the Peninsula" of Spain and Portugal?—Are there any mountains in Norway and Sweden?—Give the general direction of the mountains of Europe.—Are the mountains of Italy on the east or west side of that peninsula?—Where are the Jura Mountains?—Carpathian?—Where is Mount Etna?—Mount Vesuvius?—St. Bernard?—Mont Blanc?—How high is Mont Blanc? (See bottom of Mercator's Map of the World.)—What country is on the top of the great watershed of Europe drained by the Rhine, Rhone, Danube, and Po?—Where are the Alps?—What is the highest mountain in Europe? (Refer to the Orographic View of Central Europe.)

Rivers.

Name the rivers of France, and tell where they rise and where they empty.—(The Loire (length, 645 miles), the longest river of France is about the length of the Cumberland River in the United States.) What two rivers unite to form the Gironde?—Is any part of France drained by the Rhine?

What five rivers of Spain flow into the Atlantic Ocean?—Describe the Elbe—The Rhine—The Weser.—Trace the Vistula—The Oder—The Dwina—The Dnieper—The Volga—The Don—The Dniester—The Po—The Rhone—The Seine. *The Seine runs through the City of Paris.*—The course of the Danube.—Which watershed is the largest, that which is drained into the Baltic, or that which is drained into the North Sea?—How far is the Rhine navigable? *Ans.* to Schaffhausen, in the north of Switzerland.

NOTE.—The Elbe and Rhine drain 143,000 square miles; the Seine, Loire, and Garonne, 110,000 sq. m.; the Douro, Tagus, Gaudiana, and Guadalquiver, 115,000 sq. m.; the Thames, 6,500 sq. m.; the Severn, 5,500 sq. m.; the Neva, Vistula, Oder, Dwina, and Niemen, 305,000 sq. m.; the Ebro, Rhone, and Po, 110,000 sq. m.; the Danube, Dniester, Dnieper, and Don, 791,000 sq. m.

The Danube is, of all these, the largest, running 1000 miles in a direct line from its chief source.

Seas, Gulfs, and Bays.

Where is the Adriatic Sea?—Archipelago?—The Black Sea?—The Sea of Azov?—The Sea of Marmora?—The Caspian Sea?—The White Sea?—The Baltic?—The North Sea?—The Mediterranean? *The Mediterranean is a warm, tideless sea. The countries along its border enjoy an insular climate.*

NOTE.—The reason the temperature of the sea is not so changeable or easily affected as that of the land, is to be found in the fact that water has great capacity for heat; it can, with but slight change of temperature, hold much more heat than the air can. Thus the heat that a cubic foot of water loses in cooling 1° will warm 100° cubic feet of atmosphere 1°.

Where is the Gulf of Lions?—Of Genoa?—Name all the Gulfs on the southern coast of Europe.—Where is the Gulf of Bothnia?—Gulf of Finland?—Gulf of Riga?—Where is the Bay of Biscay?

Straits and Lakes.

Where is the Strait of Gibraltar? *The water of the Mediterranean is largely absorbed by the sun, by the process of evaporation, and, to supply this loss, there is an indraught from the Atlantic Ocean, through the Strait of Gibraltar, forming sometimes a strong current which sailing vessels can hardly stem.*

Where is the Strait of Messina? *In the Strait of Messina are the famous rocks of Scylla and the whirlpool of Charybdis, which were the dread of ancient mariners.*—Where is the Bosporus?—What does it connect?

NOTE.—In the Strait of Messina the fishery of the sword-fish is of importance. The men of Messina and Reggio join in with a great number of boats, carrying brilliant flambeaux, and use three-pronged harpoons. The swordfish is from five to six feet long, is terribly armed, and often perforates the copper and hull of ships with its weapon. A miserable little parasite, however, sometimes burrows beneath the flesh of the sword-fish, and drives it mad with pain, till it dies.

THE SWORD-FISH.

Where is Lake Lucerne.—Lake Wener?—Lake Ladoga?—Onega?

Islands and Capes.

Where are the Lofoden Isles?—The Balearic Islands?—Where is Sardinia?—Corsica?—Sicily?—Malta?—Candia?—Cyprus?—Rhodes?—What Islands are in the Baltic Sea?—The Scilly Islands?

Where is Cape Matapan?—The Naze?—Cape St. Vincent?—Cape Finisterre?—North Cape?—What is the most southern cape of Europe?—What is the most eastern?—What the most northern cape?

JAN MAYEN I.

HAMMERFE

C. North

Arctic Circle

ICELAND

Malstrom

SHETLANDS

Mainland

ATLANTIC OCEAN

BRITISH ISLES

NORTH SEA

EDINBURGH

IRELAND

LONDON

ENGLISH CHANNEL

BALTIC

STOCKHOLM

CHRISTIANIA

Heligoland

BAY OF BISCAY

FRANCE

SPAIN

PORTUGAL

LISBON

MADRID

BARCELONA

BALEARIC IS.

MINORCA

PALMA MAJORCA

IVICA

CAGLIARI

C. ST. VINCENT

STR. OF GIBRALTAR

GIBRALTAR

ALGIERS

TUNIS

MEDITERRANEAN SEA

MALTA Valletta

AFRICA

ROME

SAN MARINO

CORSICA

SARDINIA

SICILY

SYRACUSE

C. Passaro

KEFALONIA

EUROPE

Statute Miles

Harbors and Cities.

Name some of the harbors on the Mediterranean, and point them out— Also, on the Atlantic—On the Baltic and its great estuaries.—Name the Capital of every country in Europe, and state how it is situated with regard to the chief seaport of that country.—What are the chief towns of France?—Of Prussia?—of Holland?—of Belgium?—of Switzerland?—of Turkey?—of Austria?—of Italy?—of Spain?—of Portugal?—of Norway and Sweden?—of Denmark?—Point out Paris—Nijni-Novgorod—St. Petersburg—Hammerfest—Athens—Corinth—Vienna—Rome—Cronstadt—Sebastopol—Antwerp—Geneva—Amsterdam—Stockholm—Copenhagen—Berlin—Basle.

Distances, Routes of Travel, and Trade.

What town in France is opposite Dover, in England?—Distance between the two places?—How far is Paris from the sea?—How far and which way do you go from Toulouse to Calais?—To Marseille?—To Nice?—To Cherbourg?—From Calais to Marseille?—From Havre to Brest?—From Bordeaux to Nantes?—How would you go from Paris to Geneva?—How far is it from Berlin to Vienna?—Tell the distance from the Hague, by water, to Schaffhausen.—How would you go from Hamburg to St. Petersburg?—From Venice to Florence?—From Geneva to Rome?—From Athens to Constantinople?—To Sebastopol?

How far is it from Malta to Alexandria, Egypt?—From Malta to Paris?—From Bremen to Hammerfest?—From Cherbourg to Copenhagen?—From St. Petersburg to Nijni-Novgorod?—From St. Petersburg to Constantinople?—From Antwerp to Madrid?

Miscellaneous.

Can you find the Valdai Hills? *They lie between St. Petersburg and Moscow.*—The Sea of Azov?—The Straits of Bonifacio?

The Alpine peak of the Matterhorn? *It is a boundary between Switzerland and Italy.*—By referring to the text, can you point out the Mont Cenis Tunnel?—Where is the Island of Zante?—Elba?

Can you locate the Puy de Dome? (Puy means *peak*). *The Puy de Dome is in the centre of the great volcanic region of Auvergne, near Viviers, in France.* —(It was by the aid of this mountain that the two French philosophers, named Perier, first discovered that the air had weight. They found by actual experiment that the pressure of the atmosphere is *greater at the foot* than it is at the top of this mountain, and thus the barometer was invented.)

[The *Barometer* is the instrument employed to measure the height of a column of mercury supported by the pressure of the atmosphere. From this height the weight of the atmosphere is ascertained. The fundamental principle of the barometer cannot be better illustrated than by the following experiment. Take a glass tube, 34 inches in length, open at one end; fill it with mercury, and, closing the open end with the finger, invert it, and plunge the open end into a bowl also containing mercury. The column will fall in the tube to about 30 inches above the surface of the mercury in the bowl, if the experiment be made near the level of the sea. The fluid is upheld in the tube by the air outside of it pressing on the mercury in the bowl; and since the one thus balances the other, it is evident that the mercurial column will serve as an accurate indicator of the varying pressure of air. The space in the tube above the mercury is one of the nearest approaches to a vacuum that can be made. It is called the *Torricellian vacuum.*]

BAROMETER, TUBE, AND VACUUM.

Is Switzerland nearest to the Black Sea, the Mediterranean, or the Atlantic Ocean?—Where is the Peloponnesus? (See Greece.)

Where is the Hospice of St. Bernard?—By reference to the text, point out the great grain country of Europe.—How do the Germans get to the sea?—How does inland trade between St. Petersburg and Peking in China, go on? *Ans. It passes through Kiachta, the centre of the tea trade between Russia in Europe and China.—The merchandise is transported on the backs of camels from Peking to Kiachta all the way across the desert of Mongolia; and thence it is carried to Russia by river transportation in summer, and in winter by sledges drawn by dogs or reindeer.*

Review Questions.

Give some account of the political geography of Europe and its social features—Its population.—Which are the maritime nations of Europe?—Describe some of the natural peculiarities of Europe—Climates and productions—Nationalities.—Mention the chief points of interest about Paris. Lyons, Marseille, Bordeaux, Rheims, and Nantes.

[Questions from Lesson LV. to LX. are found at end of each lesson.]

LESSON LX.—*1.* Give the general geography of Austria, Hungary, and European Turkey. *2.* To what races do the inhabitants of Austria belong? —Describe the mines of Austria.—What is said of Bohemia?—Of the cities of Austria and Hungary, and their population?—The nobility? *3.* What is said of the Turkish empire and its divisions?—Of Constantinople?—Of Adrianople?—Salonica?

[NOTE.—*In mentioning large cities, give population.*]

LESSON LXI—*1.* General geography of Greece and Italy, Spain and Portugal—Their aggregate population? *2.* Cities of the Spanish peninsula? *3.* The Alhambra? *4.* Gibraltar? *5.* Alpine Passes? *6.* Mont Cenis Tunnel? *7.* Staples? *8.* What is said of Spain and Portugal?—Their cities? *9.* What is said of Italy?—What does it contain?—What is said of Rome and other Italian cities?—San Marino? *10.* Describe Greece and its cities.

LESSON LXII.—*1.* General geography of Germany and the smaller States? *2.* Hamburg and Bremen? *3.* Intellectual character of Germans? *4.* What is said of Prussia?—Berlin? *5.* What changes were wrought by the war of 1866?—German cities? *5.* What is said of Saxony? *6.* Of Belgium and its cities? *7.* Of Holland, and its people and cities?

LESSON LXIII.—*1.* What is said of Switzerland? *2.* Mont Blanc and its height?—Other peaks and glaciers?—Hospice of St. Bernard? *3.* Swiss exports?—Point out on the Orographic View of Central Europe, *without the use of the numbers*, the principal cities of Central Europe. *4.* What is said of Sweden and Norway?—Cultivation of soil?—Sun at midnight?—Industries?—Religion?—Lapland?—What is said of Hammerfest?—Its latitude?—The cities of Sweden and Norway?—Capital? *5.* What is said of Denmark?—Its climates?—Its cities?

LESSON LXIV.—*1.* What are the limits of Russia? *2.* Chief industries? *3.* "Black Lands."—*4.* Great grain-country of Europe?—Russian exports? *5.* Population of Russia and its chief cities? *6.* Steppes of Russia? *7.* Inland trade?—Fairs of Nijni-Novgorod? *8.* Russian minerals? *9.* Serfs? *10.* Moscow and other Russian cities? *11.* Government? *12.* Races? *13.* What is said of Finland?—Helsingfors?—What is the title of the chief magistrate of Finland?

FLORA AND FAUNA OF ASIA.

LESSON LXVI.

Asia. (Cap, p. ...)

1. Asiatic Races.— We now pass from the States of Europe to the oldest nations of the earth—from yellow hair, blue eyes, and fair skins, to raven locks, black eyes, and all shades of complexion ;—from the tawny and yellow to brown and black ;—from Christians to the followers of Mahomet, or Buddha, or Brahma, or Zoroaster : these are the popular sects of Asia.

DOMESTIC LIFE IN JAPAN.

2. Population.—The most populous empires in the world are in Asia, but, though they have extensive sea-coasts, skirted with beautiful islands and embellished with deep bays and capacious harbors, there is not one that has risen to importance as a maritime or naval power.

Men are respected according to their virtues, but a nation is regarded with respect by her fellows and treated with consideration, according to her physical strength and the prowess of her sons.

Without a navy, no nation can make her power felt, or spread her influence across the sea. She may make bordering nations, and those that she can reach by land, feel and confess her martial energy ; but it is quite different with those whom she has to pass the seas to reach.

Hence China, with her majestic sea-front, and a population exceeding that of all the nations of Europe combined, has never been able to command the respect abroad that is accorded even to a third-class naval power.

3. Political Geography.—Part of the domains, both of Russia and Turkey, are in Asia. Not counting these, there are in Asia but six States that are recognized as free, sovereign, and independent nations.

These are Persia, Siam, Anam, Burmah, China, and Japan. They are absolute monarchies. With their dependencies they embrace an area of 5,849,767 square miles, which is occupied by a population of 537,000,000.

The rest of this continent embraces an area

of 11,500,000 square miles, with a population of 268,600,-000, consisting partly of possessions that the maritime powers of Europe have acquired by conquest : as India, which belongs to England ; the Island of Java, and other islands, which belong to Holland ; the Philippine Islands, to Spain ; much of Cochin China, to France, etc., etc.

Or it consists of districts, such as Arabia, Turkestan, Thibet, Afghanistan, Beloochistan, etc., etc., that have no responsible government, and are divided into tribes, hordes, and other factions, which do not pretend to be clothed with the dignity of national sovereignty.

4. Modes of Life and Civilization.—Most of these hordes and factions lead a sort of half-savage, half-civilized life—tending their flocks, robbing their neighbors, and plundering the helpless.

Parts both of Asia and Africa were inhabited by civilized people long before Europe and America were. Almost all the great events recorded in the Bible took place in Asia.

5. Woman.—In these lands woman is degraded. In parts of India the practice of widow-burning is still continued ; that is, when the husband dies, his wives, for he often has many, are burned at his funeral.

6. The Grand Plateau of Asia.—If you will study the Orographic View of Central Asia (p. 133), mark where the rivers rise, and note the direction in which they flow, you will see, by the natural drainage of the land, that there is an extensive inland region which has no sea-drainage and but few watercourses.

This is the grand plateau where the Steppes of Asia are, and the great Desert of Gobi is on it. The highest land in the world is here.

Among the peaks of the Himalaya, which form the southern rim of this great inland basin, stands the majestic Gaurisankar, also called Mount Everest, towering more than 29,000 feet above the level of the sea.

—————

LESSON LXVII.

CHINA.

1. China is the oldest and the most densely populated empire in the world. With an area, including adjoining dependencies, of 4,695,000 square miles, it now contains, by my estimates, a population of 480,000,000.

The land is filled to overflowing with people. Its civilization dates from time immemorial. The country is in a high state of cultivation and im-

provement, to which Europe did not approximate, even at the beginning of the present century, and to which she has not yet attained.

2. The most famous works in China are its canal, 700 miles in length, constructed nearly one thousand years ago, and its celebrated wall, 1200 miles in length, from 15 to 30 feet high, and so thick that six men on horseback can ride abreast upon it.

It was designed as a work of defence against the Tartars, and was completed upwards of two hundred years before the birth of Christ.

3. The Chinese are an industrious, patient, economical, and ingenious people.

Of all the industrial pursuits, the cultivation of the soil is most honored by the Chinese ; and, to do it homage, the Emperor, with his nobles, and in the presence of his subjects, annually puts his hand to the plow and runs a furrow.

4. You ought now to know enough about climates and geography to tell, merely by looking at the map, that the sea-slopes of this country are well watered, and that therefore it has climates and soil adapted to the cultivation of all the great staples that are grown between the same parallels of latitude in America.

5. Tea is the *great* agricultural staple in China. It will grow equally well in our Southern States, and it is not cultivated there, simply because labor is scarce and can be more profitably employed in other branches of industry.

Tea is produced by labor in China that costs only a cent or two a day, and none of the great agricultural staples of commerce require more labor than tea. The leaves not only require to be gathered by hand, and one by one, but each one has afterward to be rolled up separately, also by hand.

They make great use of the bamboo. They build houses and bridges of it. They use it as food when it is young, and they make mats, furniture, and household utensils of it when it is matured.

They grow and manufacture silk, cotton, and calico. Every available foot of land in China is cultivated, and for the want of dwelling space, many

CHINESE RAT-CATCHER.

thousands of people live in boats, arranged in streets on the water, as houses are in a city.

The Chinese are extensively engaged in sea-fisheries, and their chief article of food is rice.

The internal commerce of China is immense, and is carried on by means of its water communication through canals and navigable rivers. In the mountain districts donkeys are used, and in crossing the deserts, the camel.

6. Their religion is Buddhism. The government is an absolute monarchy, and the laws are severe. Their country is the "Celestial Empire," and their Emperor, according to them, is of such high descent as to be brother of the sun and moon.

The geographical information of the Chinese does not extend beyond their own country, which they maintain is the centre of the world, and they have the greatest contempt for all foreigners, whom they call "outside barbarians."

7. Their domestic animals are generally of the scavenger sort, such as swine, dogs, ducks, geese, and poultry. They are very poorly off for sheep, horses, and horned cattle.

8. Their cities are compactly built and the streets are narrow. Their temples and pagodas are very grand, and enhance the beauty of many a landscape.

Rebellion, revolution, and civil war have been raging in China for more than a quarter of a century.

We have a valuable commerce with China.

Nanking, population 300,000, once an imperial city, is on the Yangtse-kiang river, or "Child of the Ocean."

Shanghai has an immense commerce and a population of 300,000.

Peking, the capital of the Empire, about 106 miles from the sea, with 1,650,000 inhabitants, is in latitude 40° north.

There are three cities in China on the Yangtse-kiang so connected with each other, that they may be said to form one city, under the name of Hoang-Chou, or Hankow, with a population of nearly 8,000,000, the largest known collection of human beings in so small a compass.

MONGOLIA

Is a part of the Chinese Empire. It is an arid and chilly country, mostly desert. The Mongols are nomadic by necessity, for their soil is too poor to sustain their flocks with grass but for a short period of time.

MONGOLIAN EMIGRANT.

17

LESSON LXVIII.

Japan.

1. The Empire of Japan consists of four large islands, viz., Niphon, Yesso, Kiusiu, and Sikoke, and 3,850 smaller islands. Its area is about one-fourth larger than that of the British islands. It is separated from Corea and the continent of Asia by the Corean Channel and the Sea of Japan: its coasts are generally bold and rocky: it has an inland sea of great beauty and abounds in convenient harbors.

2. Both these nations have opened certain of their ports, called the "treaty ports," to foreign trade and residents.

These five ports in Japan are Nagasaki, Hakodadi, Simoda, Yedo, Osaka, Hioga, Niigata, and Kanagawa.

The honor of making the first negotiations which led to the opening of Japanese ports to the commerce of the world, was reserved for Com. Perry and the naval officers of the American expedition of 1853. The first ambassadors ever sent from Japan were accredited to the United States in 1860. A large trade and very friendly intercourse have already sprung up between the two countries. Not less than five hundred Japanese students have come (1872) to the United States to be educated.

3. The Japanese trace their history by authentic records through a period of 2,539 years, under 124 emperors, who have borne the title of Tenno or Mikado.

For the last 600 years the Shiogoon or Tycoon has, as a subordinate, governed the empire; but since the revolution of 1867, the Mikado has assumed his ancient prerogatives. The influence and revenues of the formerly numerous and powerful feudal lords called Daimios, have also been greatly modified and reduced.

4. The Japanese resemble the Chinese in appearance, though they are a finer-looking race of people.

The Japanese have a literature, and writers of great antiquity and repute, and used the art of printing long before it was invented in Europe.

5. They have severe laws and singular manners and customs. When a person of rank offends the government a sword is sent to him, and he is then in honor bound to commit suicide; this is the Hara-wo-kiru.

6. Their country is volcanic. In consequence of this fact their dwellings are generally of wood, and are all built according to one of three or four plans; so that, in furnishing a house, you have only to go to the upholsterer and order carpets or mats for a house of one of these patterns.

7. They have no chairs, sofas, or beds; but, using their clothes for covering, they sleep on the floor, upon the mats on which they sit and receive their company during the day.

The married women pluck out their eyebrows, use pigments to turn their teeth black and their lips red, and powder themselves with rice flour.

RICE STOREHOUSES AT YEDO.

The usual mode of travelling is, not in carriages drawn by horses, but in either palanquins or norimons borne by two or four men.

8. The cities of Japan are numerous. Yedo, or Tokio, the present capital, has 1,550,000 inhabitants.

Miako, long known as the western capital, was founded in A.D. 794, and has 370,000 inhabitants. Osaka, the port for Miako, is accounted by its inhabitants the Paris of Japan. Yokohama is the most important port of the empire. Hakodadi is the principal port of Yesso. Nagasaki is the port where Europeans were first permitted to locate as merchants.

9. The islands of Japan lie between the parallels of 30° and 45°, and their climates resemble those of our Atlantic-seaboard, though somewhat milder, from their insular position. At Yedo, the summer temperature ranges from 70° to 90°; in winter the snow seldom lies long.

RICE CULTURE.

In addition to the staples that are cultivated with us they have the *lacquer-tree*, from which they get the gum for their beautiful Japan-ware; the wax-tree, from which they get the resin for their candles—the manufacture of which is an important branch of industry; and the *paper-mulberry*, from which they manufacture their paper.

Of the food-plants, rice is most extensively cultivated—on the hill-sides as well as in low, marshy regions, as in South Carolina and Georgia.

10. Japan also, like China, is rich in mineral resources. Both the base and precious metals are found, and coal is obtained in large quantities.

Mineral springs, both cold and hot, abound, and near the island of Kiusiu there is a small islet with solfataras and a burning volcano, which answers capitally for the mariner all the purposes of a first-class lighthouse.

Foosiyama, the Parnassus of Japan, with its majestic cone of snow, is in sight from the capital. It is an extinct volcano, 14,177 feet above the sea, and is exceedingly grand and beautiful. It is an object of veneration with the lower orders.

11. The Japanese are intelligent, industrious, and ingenious. Their porcelain and steel manufactures are equal to any in the world; their silk and other fabrics are of superior excellence and beauty; and numerous varieties of paper are made by them, many of which are very beautiful and perfect.

In the commercial and manufacturing way, the Japanese hold the same relation to the neighboring continent, Asia, that the English once held to Europe.

12. The opening of the ports of Japan and China, and the establishment of lines of steamships between them and American ports on the Pacific, are rendering important service in developing a new civilization among these peoples.

The Japanese Islands have a dense population, estimated at 35,000,000.

LESSON LXIX.

The Empires of Anam and Birmah, and the Kingdom of Siam.

1. Siam, Anam, and Burmah form part of a peninsula. They lie between the parallels of 10° and 27° north. They occupy the same geographical position, and are similar in their religious and social relations; and therefore we consider that their industrial pursuits, except as Siam may be affected by the presence of the sea, and the existence of mineral resources, are the same.

2. They are well watered, and, being mainly within the tropics, their soil is as rich, and the vegetation as rank, as light, heat, and moisture can make it.

3. In the forests here are found the *taban*, the tree which yields gutta-percha, and also the tree which yields the pigment we call gamboge. Oil-trees, from a single trunk of which many gallons of oil may be extracted, abound in Burmah. Petroleum-springs are common, and seem inexhaustible.

4. We now approach the countries where the elephant is used as a domestic animal. He is not used in China, because of the excessive population; the ground required to produce food for one elephant, would support several men.

5. The inhabitants are chiefly Buddhists in religion; but in industry and intelligence, they are very inferior to their neighbors, both on the right and left, viz., the Chinese on the one side and the Hindus on the other. Bankok is the capital of Siam; Mandelay, on the Irrawaddy, of Burmah; and Hue (*hway*), of Anam.

6. The Malay peninsula, which stretches down still farther toward the Equator, is divided into a dozen petty States, thinly inhabited and badly governed. The Malays are a piratical people, dreaded by all unarmed sailing vessels.

Singapore, an English town, is at the end of this long peninsula. It is a thriving city, with a population of 90,000. The Strait of Malacca, on which it stands, is the great thoroughfare for the sea-steamers that ply between India and China; and it is a half-way house in navigation of much importance. What Liverpool is to cotton, Singapore is to tin.

7. These countries all lie within the celebrated tin region of Asia, which embraces an area of many thousand square miles, and extends from the mountains on the north to the islands on the south, some of which, Banca and Singkep, are famous. Their tin-mines supply the markets of Europe.

8. White sandalwood, ebony, rosewood, ironwood, and the red dye-woods are all found on these two peninsulas of Siam and Malay; and here the betel-palm produces its finest fruit; the bamboo flourishes; cinnamon-trees and aromatic plants perfume the air; also the sweet-scented eaglewood, which is burned as incense in the heathen temples, is among the treasures of the forest. Indigo, cotton, the sugar-cane, rice, tobacco, and the mulberry are extensively cultivated in these countries.

9. The elephant, tiger, rhinoceros, leopard, and the buffalo are all found wild in great numbers here.

The waters teem with alligators, and the forests are alive with monkeys. With the natives, alligators are an article of food.

During the prevalence of the Southwest monsoon, the sea is blown so violently that the shores of Indo-China are inundated. The forests are then navigated. There is also an immense fall of rain, causing an overflow of the rivers, which then make short cut-offs and new channels.

Burmah is especially rich in minerals. The celebrated ruby-mines are there; they yield sapphires also.

The Burmese and Siamese profess great veneration for white animals. The white elephant is one of the dignitaries of the State. He has his palace, his minister of State, and takes rank next after the royalty.

Bankok has a population of 400,000 inhabitants, large numbers of whom live in bamboo huts afloat on rafts in the river.

LESSON LXX.

Hindostan, or India.

1. India, with its departments, including Ceylon and other islands in the Indian Ocean, constitute the East India possessions of Great Britain. They embrace an area of 1,600,000 square miles, inhabited by 195,000,000 human beings.

About one hundred years ago, what is now called British India was the seat of an empire of vast wealth and splendor.

At that time the English East India Company, having grown rich and powerful under its monopoly of trade, commenced in earnest the splendid conquest which gave to England her most valuable possessions.

2. India is now divided into the Presidencies of Bengal, Madras, and Bombay, each ruled by a British governor.

The inhabitants, exclusive of the English, who, by comparison, are few in number, are generally known as Hindus, though there is great diversity of language, manners, customs, races, and religion among them.

Though the country is still populous, it was yet more so under the native rulers of by-gone days. Its deserted capitals and decayed cities, its wonderful antiquities and splendid ruins, tell, in language the most eloquent, of departed greatness, glory, and renown. The beautiful Mogul capital, Delhi, had, in its palmy days of native rule, a population of two millions; it has now dwindled down to less than two hundred thousand.

3. The commerce of India is very large. Its exports are coffee, tea, sugar, cotton, flax, rice, tobacco, opium, indigo, hemp, gums, spices, drugs, medicines, gingelly, and almost every variety of merchandise.

The shawls of Cashmere, the muslins of Dacca, the brocades of Benares, the embroidery of Delhi, and the jewels of Golconda all figure largely in the commerce of India.

The chief seaports are Calcutta, Madras, Bombay, and Colombo.

4. India lies between the

NAVIGATION IN THE FORESTS OF INDO-CHINA.

parallels of 8° and 33° north. It is in the region of the Monsoons, so named from the Malay word for *the season*, which marks the duration of the Monsoon. For six months, including the winter, the winds come from the interior and are *dry:* these are the northeast monsoons. For the other

MONSOON IN THE HARBOR OF BOMBAY.

six months, which are the summer, the winds come from the sea and are moist. They bring clouds and make the rainy season. These are the south-west monsoons. No part of the world receives so heavy an annual rainfall as some of the hills, as Cherepungee, north of Calcutta, in India.

DIAGRAM OF THE MONSOONS.

The intense force with which, in summer, the sun strikes with his vertical rays upon the bare rocks and arid wastes of the great plateau of Central Asia,

so heats and rarifies the air there, that it rises up as from a furnace, and the air from all sides and from the distance of more than a thousand miles out to sea, rushes in to fill the vacuum. THIS MAKES THE MONSOONS.

Thus, this great plateau, this inland basin, this vast and mountainous platform that lies on the roof of Asia, presents itself in a new light. It causes the winds to blow which bring the rains that water its slopes and fill its rivers.

In the diagram, the arrows with half-barb show winds which blow for half the year.

5. The cotton of India, owing to the peculiarities of climate, is of a short staple, and therefore of quality much inferior to our "New Orleans middlings."

6. Opium is made from the poppy, and the cultivation of this plant is an important branch of industry.

The Chinese, Japanese, and the inhabitants generally of the East India Islands, are much addicted to opium-chewing and smoking.

The entire proceeds of the tea crop are said to be insufficient to pay for the opium annually brought into China and consumed there.

7. The chief cities of India are Calcutta, Murshedabad, Patna, Benares, Cawnpoor, Delhi, Lucknow, Peshawur, Pondicherry, Bombay, and Madras.

Calcutta, with a population of 1,000,000, is the residence of the Governor-general of India. Its annual exports amount to nearly $100,000,000.

Bombay has a population of 820,000, and is noted for its splendid harbor, from which it derives its name.

Madras, the capital of the Presidency of Madras, as Bombay is of Bombay Presidency, has 450,000 inhabitants. It is most unfortunately situated for trade, being exposed to violent winds, which greatly endanger its shipping.

During heavy gales on the coast of Madras, the surf breaks in nine fathoms of water, at the distance of four or five miles from the shore. The stoutest boat cannot live in it, and the largest vessels cut their cables and put to sea. So awful is the gale, at times, that the waves are smoothed and levelled down by its force, and their crests are scattered in a shower of spray, called by sailors "*spoon drift.*"

Murshedabad manufactures silks, carpets, and embroidery.

Patna is a city of mud huts, and is the emporium of the trade in opium, indigo, rice, sugar, and saltpetre. Colombo, with a population of 40,000, is the chief seaport and capital of Ceylon.

Gaya, as the birthplace of the founder of Buddhism, and Juggernaut, on account of its temple of Vishnu, are places of Hindu pilgrimage.

[List of places marked on the Orographic View of Central Asia, p. 136

1. Lassa.	8. Hyderabad.	15. Peking.
2,2,2. Desert of Gobi.	9. Samarcand.	16. Kiachta.
3,3,3. Cashmere.	10. Bokhara.	17. Canton.
4. Mt. Dhawalaghiri.	11. Tobolsk.	18. Shanghai.
5. Mt. Gaurisankar.	12. Irkutsk.	19, 19, 19. Nan-Shan Mts.
6. Mt. Kunchinjinga.	13. Yablonoi Mts.	20. Cabool.
7. Multan	14. Lake Tengri-Nor.	21. Cherapungee.

W, W, W, W. Great Wall of China.

NOTE.—Let the pupil diligently verify each of these places.]

OROGRAPHIC VIEW
OF
CENTRAL ASIA

8. Benares, with 600,000 inhabitants, with its thousand Hindu temples and its 333 mosques, the splendid one of Aurungzebe among them, and its baths, is celebrated for its diamond-dealers and the wealth of its bankers, and is a place of extensive industries and much trade. Allahabad (God's House) is another place of pilgrimage for this singular people.

9. The Jumna and the Ganges are sacred rivers, and they meet at Allahabad, where pilgrims go to bathe.

Cawnpoor and Mirzapoor are near together, and are in a fine cotton, grain, indigo, tea, and tobacco country.

Agra is noted for its mausoleum and pearl mosque.

Delhi, the capital of the ancient Mogul sovereignty, is magnificent with its ruins. It has a large trade now in jewelry and cotton goods.

Hurdwar, not far off, in the gorge of the mountains through which the sacred river runs, is *the* place of pilgrimage and fairs.

10. Lucknow, with its monuments and domes, its airy palaces, and picturesque style of architecture, is a most fairy-like city, with 300,000 inhabitants.

In quelling the Indian mutiny of 1857, the heroic Havelock made it famous in Anglo-Indian history.

Cashmere is noted for its shawls and its goats, its flowers, roses, and floating gardens. There are no roads in this valley nor any wheeled vehicles

Hyderabad is renowned for the skill of its lapidaries. It is in the vicinity of the rock-temples and monolithic palaces of Ellora, all cut out of the solid rock as it lies in the mountain.

LESSON LXXI.

Arabia, Persia, Beloochistan, Afghanistan, and Turkestan.

1. You observe, by a glance at the map and at the Orographic View, that the watershed of India is separated from the streams in these countries by the Hala Mountains, and that all of Asia west of that, and south of 40°, including Turkestan, is poorly watered.

2. The people in these dry countries are all Mahommedans. War seems to be the normal state, except in Persia and Asiatic Turkey and Russia.

Many of them live in tents, and their chief wealth consists in their flocks and herds. A man there who has as many as 1000 sheep is rich.

The annual pilgrimage to the tomb of Mahomet in Mecca, is a time for trade and traffic, and a great fair is held in the city at that time.

3. There is a marked contrast between the physical aspects of the country east, and the country west of the Hala Mountains. On the east you have the elephant, the tiger, and the monkey, the bamboo, and the banyan ; and on the west, the oak, the ash, the fig, the date, camels, dromedaries, horses, goats, cows, and sheep.

Where there is so little rain the air is dry, and radiation of heat goes on much more rapidly through a dry atmosphere than it does through a moist one; for this reason the climates of the two sides of the Hala range, even in the same latitude, and at the same elevations, are very different.

On the *dry* side the days are warm and the nights are cool, with the summers much hotter, and winters much colder than they are on the India side. *Remember this rule ;* it is an important one in Physical Geography.

ARABIA

8. (Area 1,026,000 square miles, population 4,000,-000) is celebrated for its horses, its camels, and its coffee. It is divided into a number of petty States and Provinces under separate chieftains.

Travelling in Arabia is both difficult and dangerous. We know scarcely more about the interior of it than we do about the geography of the polar regions of the north.

This part of Asia, as an inspection of any rain map will show, is as dry and sterile as the Great Desert of Sahara.

It is of interest to us now, chiefly on account of Bible associations. Mount Sinai is in Arabia.

CROSSING A DESERT.

BELOOCHISTAN,

5. With an area of 165,828 square miles, and a population of 2,000,000 inhabitants, is occupied by a number of semi-barbarous tribes who have no common ruler. Kelat, with a population of 12,000, is the chief town.

AFGHANISTAN

6. Has an area of 258,520 square miles, and a population of 4,000,000, and is inhabited by a warlike, brave, and fine-looking race of people.

They are the people who, in 1842, drove the English out of Cabool with such terrible disaster. They, too, are divided into factions or tribes, each with its separate chief, khan, or sheik.

Cabool, with a population of 60,000; Kandabar, of 75,000; and Herat, of 45,000, are its chief towns.

TURKESTAN

7. Has an area of 640,436 square miles, and a population of 7,870,000, and is, as you might infer from the map and Orographic View, for the most part a desert country. The inhabitants live now as they were said to have lived more than one thousand years ago. "They exercise robbery and live by spoil."

Bokhara—population 125,000—"the Treasury of Science," within a mud wall 24 feet high and 8 miles round, has 103 colleges and 10,000 students.

Nearly all of Turkestan is under Russian control.

PERSIA,

8. With an area of 562,326 square miles, and a population of 5,000,000, is a dry country; but wherever there is "the scent of water" the little hills rejoice on every side, the pastures are covered with flocks

and herds, and the valleys are clothed with waving corn or with the most fragrant roses.

This is one of the oldest monarchies in the world. It has the signs of decay and the marks of better days.

The artisans of Persia are skilled in various branches of industry, especially in the manufacture of silks, shawls, carpets, and small arms.

In all dry countries like these, the fruits and melons, such as grapes, pears, peaches, apricots, nectarines, cantelopes, watermelons, plums, cherries, damascenes, figs, pomegranates, etc., etc., are of unsurpassed beauty and flavor.

Teheran, the capital, with a population of 100,000, is in a region which answers in latitude and geographical aspects more nearly to Albuquerque, in New Mexico, than to any other town in the United States.

Ispahan, with a population of 180,000, with Casbin, Astrabad, and Tabriz, population 150,000, are among the chief towns.

The King of Persia is called the Sheik. He is a Mahommedan.

LESSON LXXII.

Asiatic Turkey and Asiatic Russia.

ASIATIC TURKEY,

Which is inhabited by 13,000,000 Mohammedans, 3,000,000 Christians, and 1,000,000 Jews and Gypsies, is likewise a dry country. We have some commerce with it through Smyrna, a city having a pop. of 160,000.

JERUSALEM.

SPONGE FISHERIES ON THE COAST OF SYRIA.

The Tartars are a race spread over all parts of Central Asia, chiefly in Caucasia and in the Crimea of Southern Russia.

6. SIBERIA occupies an area of 5,500,000 square miles, one-third more than the entire surface of Europe, while its population does not equal that of Scotland. The climate is intensely cold, and the mercury is frozen for several months.

The silver and other mines of Siberia are worked chiefly by exiles who have been banished from the European domains of Russia, and who are sometimes sent in vast numbers to Siberia. Poland has supplied many of these unfortunate exiles.

The *Samoiedes*, a race similar to our Esquimaux, live on the marshy shores of the Arctic Ocean. These lands are called Tundras.

Tobolsk, on the Irtish river, has a population of 22,000.

2. SYRIA, one of the provinces of Turkey, contains the land of Palestine, famous in all time for the events recorded in Holy Scripture.

Jerusalem, the Valley of the Jordan, and the Dead Sea, are visited by all Oriental travellers. The "Holy Places" are in Jerusalem.

Sponge and coral fisheries on the coast of the Mediterranean are important sources of Syrian commerce.

Sponges are found on the bottom of the sea. They are animal productions, living in water from five to twenty-five fathoms deep, growing on the rocks or on marine vegetables, and sometimes on sea shells and corals.

ASIATIC RUSSIA

3. Embraces a large portion of Asia. It includes Russian Armenia, Shirvan, which extends along the southwestern shores of the Caspian Sea, Georgia, and Siberia.

RUSSIAN ARMENIA contains Mount Ararat, where the Russian, Persian, and Turkish empires meet.

4. SHIRVAN was the scene of a bold but ineffective campaign of Peter the Great. It is famous for its springs of naphtha, an inflammable fluid, which often ignites, and, flowing into the Caspian, sets its waters on fire. Near these springs is the celebrated Field of Fire. A natural and inflammable gas issues constantly from holes in the ground, and the Guebres or Parsees, fire-worshippers, at Bakou, the chief town of Shirvan, have built their temples over the openings in the earth, and conduct the gas to chimneys in the roof of the temple, where, night and day, it burns with dazzling brilliancy.

5. GEORGIA, on the south side of the Caucasian Mountains, also belongs to Russia.

It is celebrated for the beauty of its women, with whom the Grand Turk stocks his harems. Tiflis, having a population of 40,000, is the chief town.

YOUNG TARTAR NOBLE.

PETROPAULOWSKI.

Irkutsk is the seat of government, with a population of 20,000.

Yakutsk, 5,000 miles distant from St. Petersburg, is near the Asiatic Pole of greatest cold. It is surrounded by forests and marshes.

Petropaulowski (the Port of Peter and Paul), on the east coast of Kamtchatka, is the Russian naval and military head-quarters in Kamtchatka. It is nestled in a pleasant nook, and has a fine harbor.

LESSON LXXIII.

Studies on the Map of Asia.

Boundaries.

Between what parallels of latitude and meridians of longitude does the continent of Asia lie?—How many peninsular projections do you count in Asia?—What are its natural boundaries?—What are its political boundaries on the west?—Name the great seas and bays that indent the continent.— What sea makes the deepest and widest indentation?

Bound the Chinese empire, as far as possible, by its natural or physical boundaries.—What States are its political boundaries?—What are the boundaries of Thibet?—Bound Russia in Asia, physically and politically.—Bound Hindostan—Burmah—Siam—Anam—Persia—Afghanistan—Beloochistan—Arabia—Turkey in Asia.

18

What are the marine boundaries of the Japan Empire?—Where is the Kirghis Steppe? — Kamtchatka? — Bound Corea—Anam—Siam—Bound Syria—Tartary—Mongolia—Manchooria.

Mountains, Table-Lands, and Steppes.

How do the Asiatic mountains generally run? *Ans. Not as the American mountains, from north to south, but in an eastwardly and westwardly direction.*—Begin at the Taurus Mountains, in Turkey, and trace the mountain system of Asia to the northwest of Siberia.—Describe the Altai—The Yablonoi—Thian Shan—Peling and Meling—The BolorTagh—The Karakorum—The Hala—The Himalaya—The Eastern Ghauts—The Western Ghauts—The Taurus—Caucasus—Hindoo Koosh—The Suleiman Range—Kuen Lun—The Ural Mountains—The Nanling Mountains.—Trace the limits of the table-land of Thibet—That of the Deccan in India. (See Orographic View of Central Asia)—Where is Mount Everest?

Rivers, Lakes, and Inland Seas.

What rivers of Asia empty into the Indian Ocean?—What, into the China Sea?—What, into the Yellow Sea and Sea of Japan?—What, into the Arctic Ocean?—What rivers have no sea-drainage, but are inland?—Is the Volga in Europe or Asia?—The Ural river?—The Amu. *The ancient name of this was Oxus.*

What river of Asia crosses the greatest number of degrees of latitude?—What river crosses the greatest number of degrees of longitude?—Excepting the rivers emptying into the Arctic Ocean, which is the longest river of Asia? (See at bottom of Mercator's Map of the World.)

Where is the Indus?—The Songko?—The Nerbudda?—Irrawaddy?—The Brahmapootra?—The Godavery?—The Ganges?—The Hoogly?—The Canton?—The Cambodia?—The Yang-tse-Kiang?—The Hoang Ho?—The Amoor?—The Obi?—The Yenisei?—The Lena?—The Irtish?

Where is the Caspian Sea? *This inland sea has no outlet, and, though it receives the Volga, the largest river of Europe, its level is falling, owing to the great solar evaporation from its surface.*—Where is the Aral Sea? *This and the Caspian are both salt-water lakes.*—Where is the Dead Sea?—Lake Balkash?—Lake Baikal? *Lake Baikal is the largest fresh-water lake in Siberia. In winter it is covered with ice four feet thick, and is then traversed by sledges laden with tea, from China. It receives 160 rivers and 'reams.*

Capes, Bays, Gulfs, and Straits.

Where is Cape Itas-al-had?—Cape Comorin?—Cape Romania?—Cape Lopatka?—East Cape?—Cape Engano?

Where is the Bay of Bengal?

Where is the Persian Gulf?—The Gulf of Oman?—The White Sea?—Gulf of Martaban?—Gulf of Siam?—Gulf of Tonquin?—Gulf of Cutch?—Gulf of Anadir?

Where is Amur Gulf?—Strait of Malacca?—Strait of Formosa? *On the east side of the island of Formosa the Black-stream of Japan sets very strongly to the north; on the west side of the island the cold counter-current from the Arctic seas sets strongly to the south. A branch of it enters the seas between Japan and China through the Gulf of Tartary, the Straits of Perouse and Sangar, and the Corea Channel.*—Where is Behring Strait?

ASIA

Statute Miles.

MEDITERRANEAN

AFRICA

EGYPT

NUBIA

ARABIA

RED SEA

TURKEY

PERSIA

AFGHANISTAN

BELOOCHISTAN

TURKESTAN

KIRGHIS STEPPE

RUSSIA

CHINESE

TIBET

HINDOSTAN

DECCAN

ARABIAN SEA

INDIAN OCEAN

BAY OF BENGAL

BLACK SEA

PERSIAN GULF

GULF OF ARABIA

Equator

Tropic of Cancer

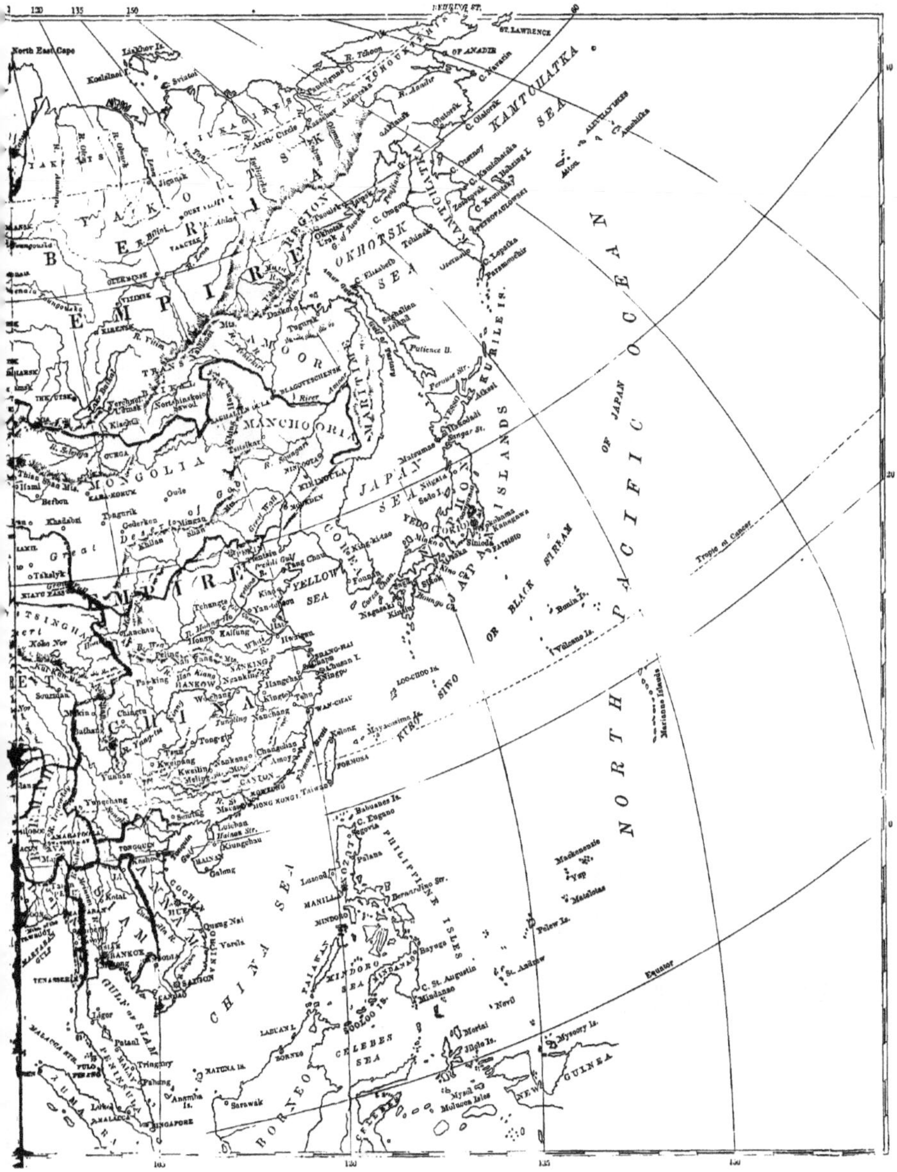

North East Cape

BERING SEA

OKHOTSK SEA

KAMTCHATKA

ST. LAWRENCE

ARCTIC CIRCLE

YAKOUTSK

BERING EMPIRE

TRANS

MANCHOORIA

MONGOLIA

KURILES IS.

SEA OF JAPAN

JAPAN

YELLOW SEA

CHINA

TSINGHAI

NAN-KING

HANKOW

CANTON

HONG KONG

FORMOSA

LOO-CHOO IS.

KYRU

SIWO OR BLACK STREAM

OR JAPAN

NORTH PACIFIC OCEAN

Tropic of Cancer

Marianne Islands

COCHIN CHINA

SAIGON

MANILA

PHILIPPINE ISLANDS

LUZON

MINDORO

PALAWAN

MINDANAO

CHINA SEA

GULF OF SIAM

TENASSERIM

SINGAPORE

BORNEO

CELEBES SEA

SARAWAK

Moluccas Isles

NEW GUINEA

Equator

Seas and Islands, and Deserts.

Where is the Arabian Sea?—The Red Sea?—The Caspian Sea?—The Aral Sea?—The China Sea? *This sea is often swept by the fearful storms known as typhoons.*—Where is the Yellow Sea?—Sea of Okhotsk?—Behring Sea?

How many Japanese islands are there?—Name four.—Where is the island of Socotra?—Where are the Laccadive Islands?—The Maldives?—Ceylon?—Nicobar?—Andaman Islands?—Hainan?—Hong-Kong Island?—Saghalien?—Staten Island?—Kurile Islands?

Trace the limits of the Arabian Desert—The Great Desert of Gobi, or Shamo. *Gobi means "naked desert," a term characteristic of this desert, which is covered with loose sand, bare rock, shingly stones, and water-worn pebbles.*

Cities, Routes, and Distances.

Where is Jerusalem?—Damascus?—Mecca?—Mocha? *Our best coffee comes from Mocha.*—Teheran?—Ispahan?—Bokhara?—Cabool?—Tobolsk?—Yakutsk? *This is near "the Asiatic pole of Greatest Cold."*—Calcutta?—Bombay?—Singapore?—Hong-Kong?—Canton?—Shanghai?—Peking?—Where is Osaka?—Nagasaki?—Yokohama? *This is the port of Yedo.*—Hakodadi?—Petropaulowski?—Point de Galle?—Bankok?—Yedo (Tokio)?

How would you go from Aden to Bombay?—From Bombay to Calcutta?—Point de Galle to Singapore?—Shanghai to Hakodadi?—From Peking to St. Petersburg? (See page 126, at the end of Map Studies on Europe.)

How would you go from Yakutsk to St. Petersburg?—What is the distance in a straight line from Peking to Yakutsk?—From Canton to Hakodadi?—By water from Hong-Kong to Singapore?—From Calcutta to Cashmere?—From Jerusalem to the mouth of the Indus?

Miscellaneous.

Find Mount Sinai.—Where was ancient Babylon?—Point out the Tigris and Euphrates rivers.—How far is Bombay from Calcutta?—Find Lucknow—Delhi.—What is Thibet? *Ans. A plateau. In Thibet and Tartary is found the strange Yak, or grunting ox, which loves to roam above the snow-line.*

THE YAK OF TARTARY.

Where is the great watershed of Asia? *Ans. N. of the Kuen-lun Mts. ; this watershed is called "the Roof of the World."*—Measuring by the scale, what is the length and breadth of the Desert of Gobi? *This is the cradle of the monsoons of the Indian Ocean. Study here the Orographic View of Central Asia.*

The southeastern portions of Asia and the northeastern regions of Africa are subject to peculiar dust-whirlwinds, which bear, in miniature, a resemblance to cyclones or revolving storms. In the northern hemisphere they revolve against the hands of a watch; in the southern, with these hands.

A DUST-WHIRLWIND.

Where is the volcano of Foosiyama?—Where are the Bonin Islands?—What are the *Tundras? These are vast marshy plains on the Arctic Ocean.*

Where is the Dead Sea? *It is 1300 feet lower than the Mediterranean.*

Review Questions.

LESSON LXVI.—*1.* What do you know of the Asiatic races? *2.* Of the population of Asia? *3.* Political geography? *4.* Its modes of life and civilization? *5.* Woman in Asia? *6.* The grand plateau of Asia?

LESSON LXVII.—*1.* Describe China. *2.* Its famous works. *3.* What is said of the Chinese? *4.* For what productions are their soil and climates suited? *5.* Name the great agricultural staples of China, and describe their uses. *6.* What is said of the Chinese religion? *7.* Domestic animals? *8.* Name the cities, and give their population.—What is said of Mongolia?

LESSON LXVIII.—*1.* Describe the Japanese. *2.* What are their free ports, and who opened them? *3.* Explain their government. *4.* Their literature *5.* Their laws. *6.* The nature of their country. *7.* Domestic furniture. *8.* What is said of the Japan Islands? *9.* Their climates and products? *10.* Minerals? *11.* Japanese character? *12.* Cities?

LESSON LXIX.—*1.* Describe Siam, Anam, and Burmah. *2.* Soil. *3.* Forests. *4.* What is said of the elephant? *5.* Trade and religion? *6.* The Malay Peninsula?—Singapore? *7.* Tin region of Asia? *8.* Products? *9.* Animals?—The white elephant?—Bankok and its population?

LESSON LXX.—*1.* Extent, history, etc., of India? *2.* Divisions? *3.* Commerce? *4.* Seaports?—Explain the Monsoons. *5.* Cotton of India? *6.* Opium? *7.* Cities? *8.* Describe Benares. *9.* Describe other cities of India. *10.* Lucknow—Vale of Cashmere—Hyderabad.

LESSON LXXI.—*1.* Describe the countries of this chapter. *2.* People. *3.* Physical aspects.—Mention a rule of physical geography. *4.* Describe Arabia. *5.* Beloochistan. *6.* Afghanistan. *7.* Turkestan. *8.* Persia.

LESSON LXXII.—*1.* Describe Asiatic Turkey. *2.* Syria. *3.* Asiatic Russia—Russian Armenia. *4.* Shirvan. *5.* Georgia. *6.* Siberia—Petropaulowski.

AFRICAN FAUNA AND FLORA.

LESSON LXXIV.

Africa. (Map, p. 144.)

1. General Geography.—Africa is to the geographer an unknown land. It is the abode of the negro

TYPE OF A CIVILIZED NEGRESS.

and of the wild Arab. Out of its vast forests and impenetrable jungles come forth those strange animals, the gorilla and chimpanzee, that bear such a painful resemblance to the human form.

2. Political Geography.—Egypt and the Barbary States profess the Mahommedan religion, and own a certain degree of allegiance to Turkey.

Algeria and St. Louis belong to France. The Cape of Good Hope, Natal, the Diamond District, and Sierra Leone, are English colonies.

Our telescopes have made us better acquainted with the geography of the moon than exploration has yet made us acquainted with the interior of Africa.

3. Deserts.—The Libyan Desert, Nubia, and the Great Desert of Sahara are attracting great interest among physical geographers.

You observe that they are situated near the Tropic of Cancer where the noon-day sun is vertical in midsummer. The atmosphere there is singularly dry and clear, and the sun is said to beat down through it with terrific force, so much so that a traveller there has called the sand " fire, and the air flame." Notwithstanding this heat by day (120° Fahr.), the nights were so cold that he often found the water in his canteen frozen in the morning.

4. Rivers.—There are streams of water here and there, such as the Nile and the Niger, that flow through this country and spread fertility as far as their waters can be conveyed by artificial means.

5. Elevations.—The hot sun in the desert so rarefies the air that it draws the winds in from the sea; they come loaded with moisture, of which the mountains rob them as they pass. Thus the Nile is fed, and a large portion of the country made fruitful, which but for the mountains would be desert too.

6. Seasons.—In the intertropical countries of Africa,

as in those of Asia and America, the year is divided into the rainy season and the dry.

In all intertropical countries of the Northern Hemisphere, the rainy season commences in the summer-time.

WHITE NILE.

Remember this; it is a physical fact and a geographical law of great importance.

7. The Nile and other Rivers.—In the rainy season, the head-streams of the Nile are flooded. It is a long river, and it is autumn before the floods get down to Egypt.

It has been discovered that the sources of the Nile extend at least as far as Lake Victoria Nyanza, in latitude 3° south.

This makes the Nile, when measured in a direct line from source to mouth, the longest river in the world.

There are two streams which form this great river: the White Nile and the Blue Nile.

The Zambesi drains the southern declivity of the same divide which sheds its waters off to the north through the Nile.

8. The exploration of Africa does not furnish material sufficient for the construction of an accurate map of its immense territory.

9. Productions.—From the western coasts of Africa we get palm-oil, elephants' tusks, which serve as ivory, dye and ornamental woods, and gold-dust.

From the northern provinces, such as Egypt and Morocco, we get, besides cotton, dates, morocco-leather, silken and other fabrics.

Southern France and other parts of Europe are supplied with early fruits and vegetables from Algeria.

From Cape Colony we get wines, and in Natal is found everything, except naval stores, that is grown in our Gulf States. Much wool is produced in these colonies. Large diamonds are found there.

10. The Climates.—The climates of Africa are diversified.

On the north a chain of lofty mountains prevents the ingress of the cool north winds, while on the northeast the country is level, and freely admits the hot blasts from Arabia.

The western coast is bathed by a cool current from Biscay Bay on the north side of the Equator, and on the south side by a current from the Antarctic Ocean, which sometimes drifts icebergs near the Cape of Good Hope.

The eastern coast is washed by the hot gulf stream, which issues from the Indian Ocean between Madagascar and the Mozambique coast, and consequently has a climate answering to that of Florida and Georgia, opposite to our own Gulf Stream.

LESSON LXXV.

Countries of Africa.

MOROCCO, ALGERIA, TUNIS, AND TRIPOLI.

These are known as THE BARBARY STATES, and extend along the Mediterranean for 2,000 miles.

PURSUIT OF THE RHINOCEROS.

This region of country was in ancient times a dependency of Rome; it supplied the armies and navies of Carthage. Algeria and Morocco are traversed by the Atlas mountains. In the Beled-el-Jerid there are groves of the date-palm, which supply the natives with food and protect them from the sun. This country is visited by the hot winds from the Sahara. The lion, elephant, rhinoceros, camelopard, camel, and ape, a species of tailless monkey, are the chief animals.

Morocco is an independent State. Morocco, its capital, contains 100,000 inhabitants.

ALGERIA is a colony of France, with a population of 3,000,000. Algiers, the capital, has 60,000 inhabitants, and is a favorite resort for invalids in winter.

TUNIS embraces in part the territory of ancient Carthage, and now has a population of 1,000,000. Its capital, Tunis, is the largest city of the Barbary States, with 150,000 inhabitants.

TRIPOLI has only one-half the population of Algeria, and, having no mountains between it and the Sahara, is sterile, the sand of the desert being blown up to the very margin of the sea. Tripoli, the capital, is a starting-point for explorers of the Sahara.

The palm is a tree of great value in this part of Africa.

LIBERIA.

2. Liberia is an American settlement of emancipated negroes, established in 1823, under the auspices of the American Colonization Society. It is a small but independent republic, with 700,000 inhabitants.

3. SENEGAMBIA, UPPER AND LOWER GUINEA, lie in the west coast region of Africa, and are famous for their products of palm-oil, cotton, ginger, and gold-dust. Sierra Leone, in Senegambia, is an English asylum for recaptured Africans.

NUBIA AND KORDOFAN

4. Are under the rule of the Pacha of Egypt. In these sections of Egypt the bed of the Nile is much depressed, and water is obtained on the high banks by water-wheels. Khartoum, near the confluence of the Blue and White Nile, is a city of 40,000 inhabitants.

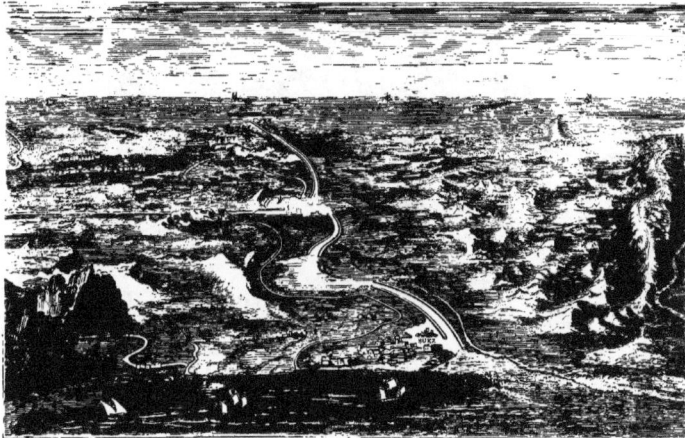
PALM-TREE.

EGYPT.

5. In olden times Egypt was the granary of the world. It is now governed by a Viceroy of the Sultan

PANORAMIC VIEW OF THE SUEZ CANAL.

M. Lesseps, a Frenchman, has made the present Pacha's reign glorious by constructing a canal across the Isthmus of Suez.

It has a depth of 26 feet, and was opened to commerce in the autumn of 1869. It connects Suez on the Red Sea and Port Said on the Mediterranean.

Egypt is rich in ruins and monuments of former greatness.

The Valley of the Nile, though of unrivaled fertility, lies in the rainless region. But, notwithstanding the fact that often for months not even a drop of rain falls in the Valley of the Nile, this great river rises annually and almost on the same day every year, ; its overflow enriches the soil.

Alexandria, the principal seaport of Egypt, has a population of 165,000.

Cairo, the capital, and the largest city of Africa, contains 260,000 inhabitants.

Suez, connected with Cairo by rail, and with a population of 8,000, is at the head of the Red Sea.

ABYSSINIA

6. Consists of three States, under the government of an Emperor. High table-lands mark this part of Africa, and they form the watershed of the Blue Nile.

Ankobar, the chief town in Abyssinia, with a population of 10,000, enjoys a most delightful climate.

7. THE ORANGE REPUBLICS.—In the rear of Natal lie the "Orange Republics," two small Dutch settlements; and the Diamond Fields.

8. NATAL.—The climate of Natal, and the neighboring regions along the east coast of Africa, adapts this country to the cultivation of the staple productions of Georgia and South Carolina.

9. MADAGASCAR is the largest island that anywhere curtains the shores of Africa. It is an independent kingdom.

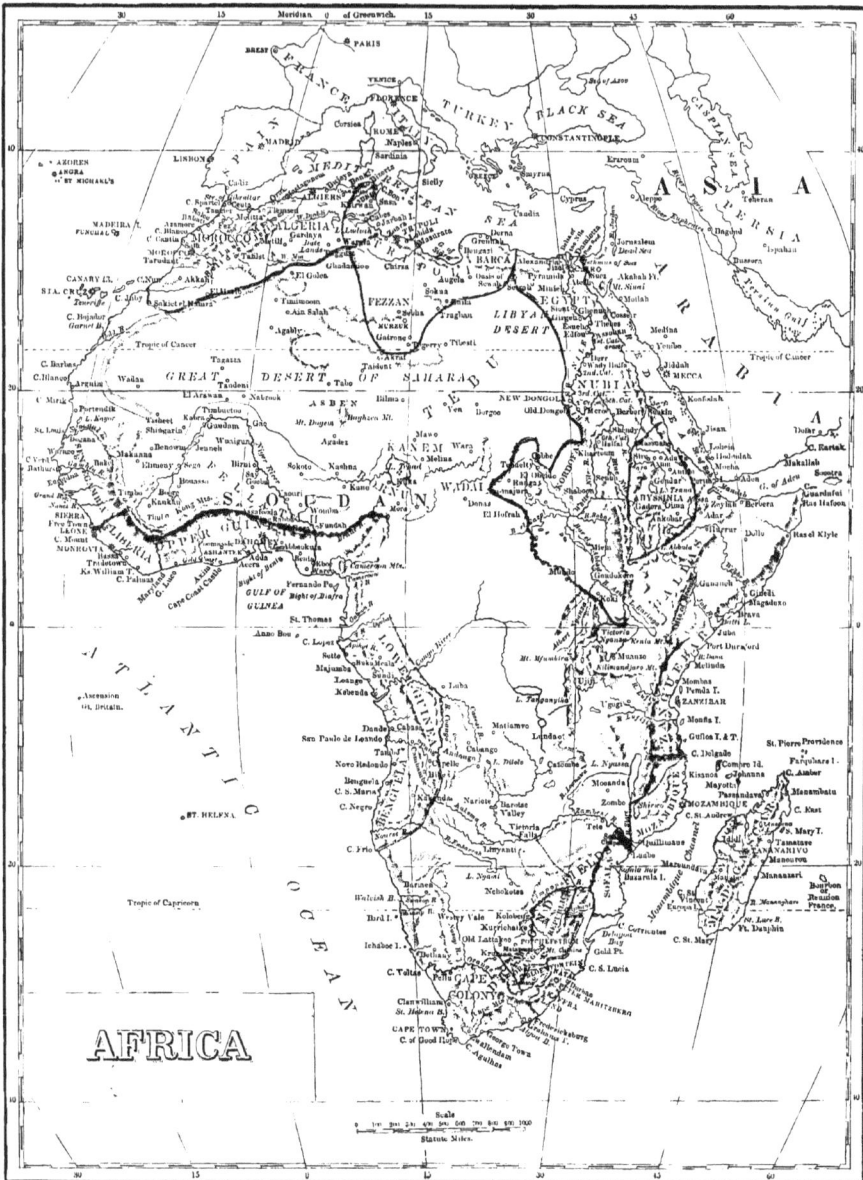

AFRICA

LESSON LXXVI.

Studies on the Map of Africa.

Boundaries, Capes, and Mountains.

In what latitude is the most northerly point of Africa?—The most southerly point?—Bound the Barbary States—Liberia—Abyssinia—Guinea—Senegambia.—Bound the Great Desert of Sahara.

What is the most northerly cape of Africa?—the most easterly?—The most westerly?—The most southerly?—Length and breadth of the Red Sea?

THE MAGELLANIC CLOUDS AND THE SOUTHERN CROSS.

A vessel on reaching the southern hemisphere, comes clearly in sight of the Magellanic clouds and the Southern Cross—the most famous objects in the southern skies. At a certain time of the year the Cross stands erct at midnight. The Indians in South America call out the time in mournful cadence, " It is past midnight, the Southern Cross begins to decline."

Where is Cape Palmas?—Cape Blanco?—Cape Lopez?—Cape Corrientes? —Cape Verd?—Cape Bon?—Where is C. Frio?—C. Guardafui?—C. Spartel? Name the principal range of mountains in the north of Africa—In the east—In the west.—Where are the Kong Mountains?—The Cameroons?— The Atlas Mts.?—The Cape Mts.?—Mts. of the Moon?—Point out the mountains in Abyssinia.—Where is Mount Kilimandjaro?—Mount Kenia? *These are lofty peaks nearly under the Equator, perpetually clad in snow ; the former is 21,000 feet high.*

Bays, Gulfs, Lakes, and Rivers.

Where is Sofala Bay?—Algoa Bay?—Delagoa Bay?—Gulf of Guinea?— Gulf of Sidra?—Gulf of Aden?—Bight of Benin? *Bight is a bend in the seacoast, making an open space like a bay.*—Bight of Biafra?

Where is Lake Tchad?—Where is Lake Tanganyika? *This lake was discovered in 1859 by Captains Burton and Speke, and partially explored in two canoes.*

Where is the Gaboon river?—The Zambesi?—The Limpopo?—Where is the Niger river?—What separates its headwaters from the sea?

Describe the Orange river—The Congo—The Gambia—The Senegal— The Nile.—Point out the White Nile—The Blue Nile—Their Cataracts.

Islands, African Cities, and Oases of Sahara.

Where is Ascension Island?—To whom does it belong?—Describe Madagascar.—What island lies east of it, and to whom does it belong?—Where are the Cape de Verd I'ds? (see p. 21.)—The Madeira?—The Azores?—The Canaries?—What celebrated peak in the Canaries? *Ans. The Peak of Teneriffe, a volcano seen in eruption by Columbus on his first voyage to the New World, 12,182 feet high. The winds at the top and bottom of the peak often blow in contrary directions. These are the upper and lower currents of the trade-wind.*

Where is Cairo?—Alexandria?—Khartoum?—Ankobar?—Mozambique? —Tananarivo?—Petermaritzberg?—Natal?—Cape Town?—Georgetown?

—Benin?—Free Town?—Monrovia?—Timbuctoo?—Fez?—Algiers?— Tunis?—Tripoli?—Find Timimoon, Murzuk, and Tegherry, in the Desert of Sahara. *This Desert is a waste of sand, like the upheaved bottom of a great sea.*

Miscellaneous.

Where is the Mozambique Channel?—What current sweeps through it? —Capital of Madagascar?—What is the capital of the Madeira Islands?— Where is Santa Cruz?—Angra?—The Gallas country?—What isthmus unites Asia and Africa?—Where is the Island of St. Helena?—Where are the Pyramids? *Ans. Near Cairo.*—The ruins of Thebes? *Ans. On the Nile, in latitude 25 30'.*

Here is a scene off the Cape of Good Hope, there the waves run so high as to hide one vessel from another even when they are close by. This Cape and Cape Horn are called the Storm Capes.

BIG WAVES OFF THE CAPE OF GOOD HOPE.

Review Questions.

LESSON LXXIV.—*1.* For what is Africa remarkable? *2.* Its political geography? *3.* Describe its deserts. *4.* Rivers. *5.* Elevations. *6.* Seasons.—What important geographical law is mentioned in this connection? *7.* Describe the Nile. *8.* What can you say of African explorations? *9.* Productions? *10.* Climates?—What warm Stream washes the eastern coast of Africa?

LESSON LXXV.—*1.* Name the Barbary States.—How far do they extend? —What is said of this region in ancient times?—Describe Morocco—Algeria --Tunis—Tripoli.

2. Describe Liberia—Its population. *3.* Senegambia, Upper and Lower Guinea.—Their productions—Sierra Leone.

4. Under what authority are Nubia and Kordofan?—How is water obtained on the Nile where the banks are high?—What is said of Khartoum? —Its population?

5. Describe Egypt.—The constructor of the Suez Canal.—Describe the work.—What points does it unite?—What is said of the Valley of the Nile? —Time of the river's overflow?—Alexandria?—Cairo?—Suez?

6. What is said of Abyssinia?—The watershed of the Blue Nile? *7.* The Orange Republics? *8.* What is said of the climates of Natal? *9.* Madagascar?

19

LESSON LXXVII.

Australia. (Map, p. 148.)

1. Area and Colonies.—This is the largest island in the world. It is owned exclusively by Great Britain. She has established in it five colonies, leaving room for a sixth. She has established colonies also in Tasmania and New Zealand.

These seven colonies are New South Wales, West Australia, South Australia, Victoria, Queensland, Tasmania, New Zealand.

2. Antipodal Relations.—Australia is on the side of the world opposite to us, with opposite seasons. There the sun casts its shadow to the south at noon, and there Christmas comes in midsummer. It is the very opposite in several other respects also.

There the leaves of the trees are not green, but dull brown, or leaden gray. The sun is so hot and the air so dry that the narrow leaves arrange themselves vertically instead of horizontally, and both sides are alike. There a forest is seldom found. The trees group themselves in clumps; they cast shadows, but they make not much more shade than would so many leafless branches. In autumn, some of the trees shed their bark instead of their leaves.

3. Vegetation and Animals.—In Australia there are nettles with stalks nine feet in circumference and forty feet high; and there are no aboriginal quadrupeds or beasts of prey larger than the Dingo dog, a sort of wolf; there is also an opossum (the kangaroo) which lives on vegetables, and, with the assistance of its tail, runs on two legs with the speed of a race-horse. The ostrich of Australia (the emu) is six feet high.

4. Climates.—Australia is for the most part a dry country, the reason for which you will understand when you learn physical geography.

5. Natives.—The natives are a sort of negro without woolly heads, but with thick lips and flat noses. Their complexion varies from chocolate-brown to sooty-black.

6. Minerals.—These colonies are rich in other minerals besides gold, especially in copper.

7. Cities.—Brisbane, Sydney, and Melbourne, the capital of Australia, are important marts.

8. New Zealand and Tasmania.—New Zealand is also rich in gold. Considering its latitude, its climates are mild, and it is a fine agricultural country. The Maoris, a fierce race, are the aboriginal inhabitants.

These islands, being supplied with warmth and moisture by the sea-winds, will produce anything that is grown in similar latitudes in other parts of the world.

South of these islands, in the Antarctic Ocean, lies Victoria Land, discovered by Sir James Ross in 1841.

9. The Tides.—It is in the ocean south and east of Tasmania that geographers locate the birth of the great tidal-wave which affects nearly every sea-coast on the globe. The tides are regular movements of the water of the sea, which ebb and flow twice every day.

The prime cause of the tidal-wave is the attraction of the moon, or the attraction of the moon and sun in conjunction, upon the deep waters.

When the moon stands over the deep sea it causes the water to bulge up and form a wave a few inches high. This is the beginning of the tide-wave. As the earth is always turning on its axis, it presents all its meridians successively toward the moon, and thus the tide-wave is kept at constant high-water on the side nearest to the moon. (*See Diagram, next page.*)

VIEW OF SYDNEY.

The moon most strongly attracts that part of the earth *nearest* to it, and it attracts the centre of the earth more strongly than the opposite side. The water on the opposite side, therefore, remains in arrear, and *apparently* recedes, forming a second protuberance on the surface of the sea. Thus we have two *high tides*, and half-way between them we have two *low tides*. The tide-waves move with a velocity of 1000 miles an hour. They are not, however, currents like the Gulf Stream, but waves like "the waving grain."

MOUNT EREBUS, IN VICTORIA LAND.

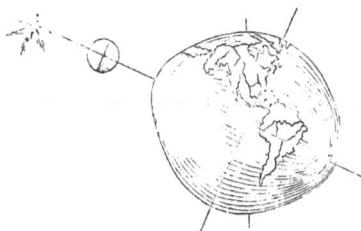

DIAGRAM OF LUNI-SOLAR TIDE.

LESSON LXXVIII.

Oceania. (Map, p. 148.)

1. Oceania lies chiefly in the Pacific Ocean, and south of the equator.

Most of its islands are either of coral formation, or madreporic, or of volcanic origin.

2. The Coralline.—The coralline is a small creature not half the size of a mite in the cheese, yet it builds up from the bottom of the sea vast islands, upon which sea-shells gather, birds light, seeds drift, and then plants grow, and after that, man comes to occupy and replenish.

3. Coral Reef.—There is a coral reef that skirts the northeast coast of Australia for more than a thousand miles

It, like all the other coral formations and sea-shells, is composed chiefly of the lime which the rains on shore dissolve, and which the rivers bring down to the sea. These little creatures, called corallines, have the power, like all shell-fish, of separating this lime from the water, of reducing it to the solid state, and of converting it into structures of various kinds.

These islands are not unfrequently surrounded by coral reefs at some distance from the shore. The great barrier reef of Australia varies, in its distance from the shore, from a few yards to 40 miles, and in its breadth nearly as much. These reefs just reach the surface of the sea and serve as breakwaters. Between them and the shore, ships ride in deep and smooth water.

4. Instincts of Corallines.—The little corallines leave gaps here and there in the reefs, so that the water may circulate freely and keep them supplied with "brick and mortar" for the vast structures that they build.

5. Offices of Sea Currents.—Thus you see that the currents of the sea perform the office of "hod-carriers" for these little masons of the deep.

The East Indies. (Maps, pp. 148 and 152.)

6. The islands of the Indies are, in eastern phraseology, "the gardens of the sun." They are intertropical, and resemble in their climate and productions our own West Indies. The East Indies are rich in gums and spices. Pop. 28,000,000 ; area 800,000 sq. miles.

NEW GUINEA (area 274,500 square miles, population 1,000,000) is inhabited by uncivilized negroes.

The bird of Paradise, whose beautiful plumage ladies often wear on their bonnets, is a native of this island.

SUMATRA, with an area of 174,170 square miles, and a population of 2,600,000, is inhabited chiefly by Malays, by whom also all of these islands, except New Guinea, are inhabited. Many of these Malays are pirates.

JAVA has an area of 52,000 square miles, and a population of 14,000,000. This island, though not the largest, is the most populous, wealthy, and influential of the East India islands. It is the "pearl of the Indies." It belongs to the Dutch, and yields a revenue of $10,000,000 annually. It exports coffee, tobacco, sugar, and various other articles.

The Islands of Banca and Billiton, famous for their tin-mines, also belong to the Dutch.

THE CELEBES, another rich group, with an area of 50,000 square miles, and a population of 300,000, are inhabited by a number of independent tribes, of whom "the Boogis" are the principal.

THE MOLUCCAS (area 70,000 square miles, population 2,000,000), are also famous for their spices. They are controlled chiefly by the Dutch, who carry on an important trade through the little island of Amboyna.

BORNEO, with an area of 289,070 square miles, and 1,200,000 inhabitants, the largest of these, is for the most part still under the rule of its dusky sons, called Dyaks. They are very warlike.

THE PHILIPPINE ISLANDS, with an area of 114,120 square miles, and with

CORAL ISLAND OF OENO AND ITS LAGOON.

a population of 6,000,000 inhabitants, belong to Spain. Next to Java these are commercially the most important of the East Indies. We have a valuable trade with them in rice, sugar, hemp, tobacco, nutmegs, and other spices.

These, with the islets and groups adjacent to them, together with Ceylon, the Maldives, and the Laccadives, the Seychelles, the isles of Bourbon and Mauritius, constitute the chief islands and groups of the East Indies.

POLYNESIA.

7. Further to the east we have *Polynesia*, or the *many islands*, which dot the Pacific Ocean through the space of many millions of square miles.

Examine the map and you will see how numerous they are, and how broad is the area over which they are sprinkled.

Here we find, on shore, in their greatest perfection, the bread-fruit, the cocoanut, and the sweet-scented sandalwood, which last is burned by the Chinese as incense in their pagodas.

Madrepores are animal flowers of the great deep, distinct from the coral, and are remarkable for the calcareous

THE BREAD-FRUIT.

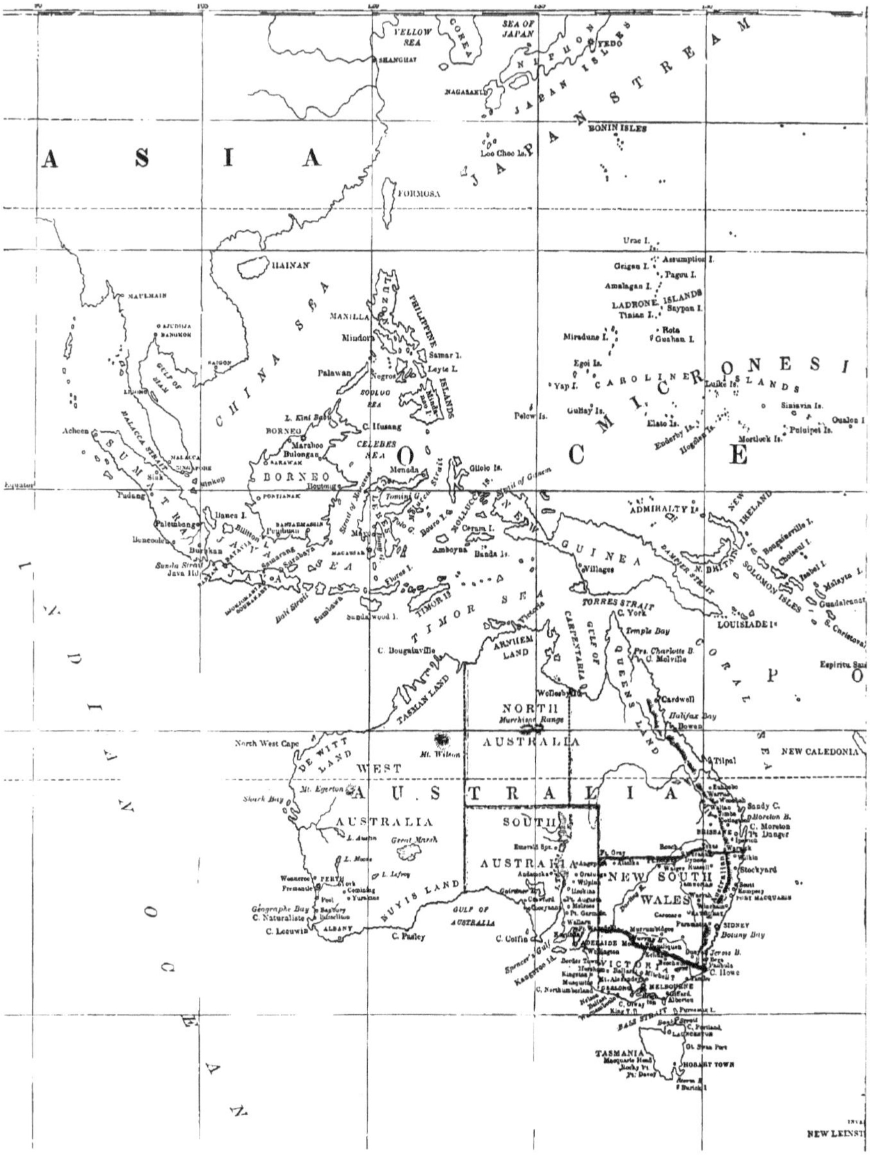

ASIA

YELLOW
SEA

SEA OF
JAPAN

COREA

NIPHON
YEDO

JAPAN STREAM

SHANGHAI

NAGASAKI

BONIN ISLES

Loo Choo Is.

FORMOSA

HAINAN

MAULMAIN

BANGKOK

MANILLA

PHILIPPINE ISLANDS

Mindoro

Samar I.

Leyle I.

Palawan

Negros

SOOLOO
SEA

CHINA SEA

SAIGON

GULF OF SIAM

Urac I.

Guagan I.   Assumption I.

Amalagan I.   Pagon I.

LADRONE ISLANDS

Tinian I.   Saypan I.

Miradune I.   Rota

Guahan I.

Egoi Is.

Yap I.   CAROLINE ISLANDS

Pelew Is.   Guliay Is.

MICRONESIA

OCEA

Luko Is.

Elate Is.   Sinavian Is.

Endorby I.   Ousloo I

Hogoleu   Mortlock Is.

Puinipet Is.

ACHEEN

L. Kini Balu

C. Huasag

SARAWAK

BORNEO

Marahoo

Bulongan

CELEBES
SEA

Menado

Gilolo Is.

Group of Guahan

BORNEO

PONTIANAK

MALACCA

SINGAPORE

SUMATRA

Palembang

Banca

Bencoolen

Sunda Strait
Java Hd.

Padang

Usnes I.

Penghuan

Sambawa

Timor G.

Boero I.

Ceram I.

Amboyna   Banda Is.

FLORES
SEA

Sandalwood I.

TIMOR   SEA

C. Dougainville

ADMIRALTY Is.

NEW GUINEA

Villages

NEW IRELAND

NEW BRITAIN

SOLOMON ISLES

Bougainville I.

Choiseul I.

Isabel I.

Malayta I.

Guadalcanar

C. Christina

LOUISIADE Is.

TORRES STRAIT

C. York

Temple Bay

Pra. Charlotte B.
C. Melville

Espiritu San

CORAL

NORTH

Murchland Range

AUSTRALIA

ARNHEM
LAND

GULF OF
CARPENTARIA

QUEENSLAND

Cardwell

Halifax Bay

Bowen

Wellesley Is.

TASMAN LAND

DE WITT
LAND

North West Cape

Mt. Wilson

WEST   AUSTRALIA

Mt. Egerton

Shark Bay

AUSTRALIA

Great Marsh

L. Austin

L. Moore

L. Lefroy

NEW CALEDONIA

Tilpal

SOUTH

AUSTRALIA

Emerald Spa.

BRISBANE

Sandy C.

Moreton B.

C. Moreton

Danger

Stockyard

NEW SOUTH

WALES

PORT MACQUARIE

WEST

AUSTRALIA

Wanneroo

PERTH

Fremantle

York

Geonin

L. Lefroy

C. Naturaliste

Gèographe Bay

ALBANY

C. Leeuwin

C. Pasley

BUYIS LAND

GULF OF
AUSTRALIA

Spencer G.

Kangaroo I.

C. Coffin

GULF ST VINCENT

ADELAIDE

VICTORIA

MELBOURNE

Geelong

SYDNEY

Botany Bay

C. Howe

BASS STRAIT

C. Portland

TASMANIA

HOBART TOWN

Pt. Davey

NEW ZEALAN

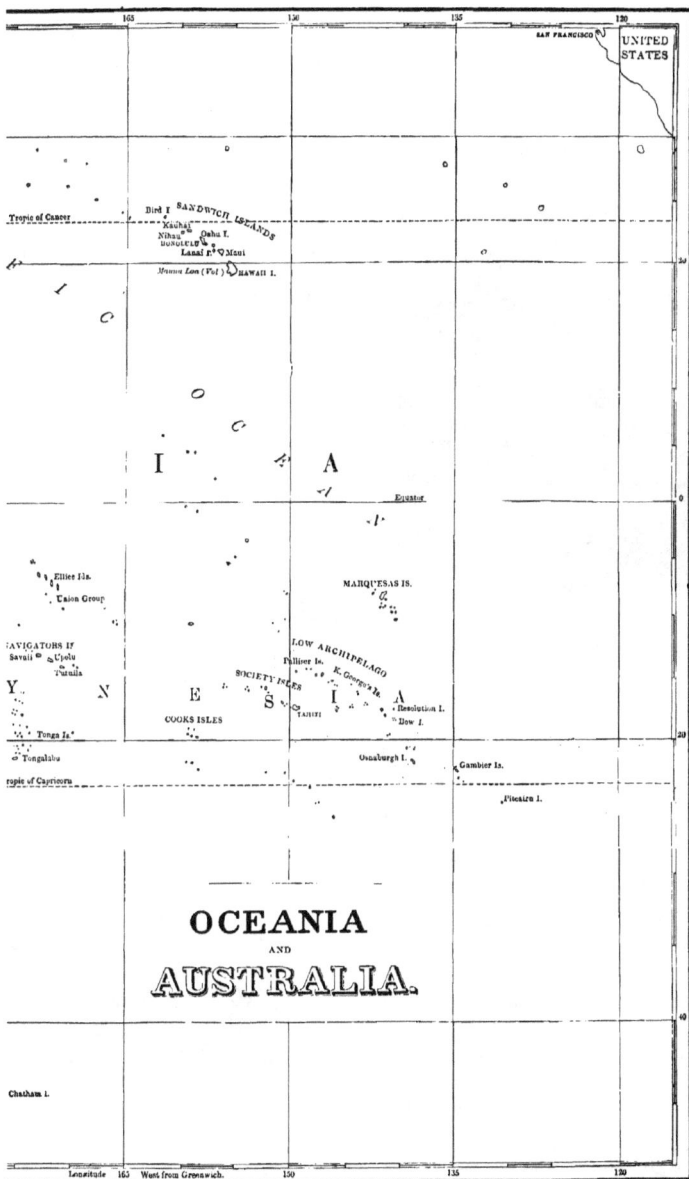

SAN FRANCISCO

UNITED STATES

Tropic of Cancer

Bird I   SANDWICH ISLANDS
Kauhai
Nihau   Oahu I.
HONOLULU
Lanai I.   Maui
Mauna Loa (Vol.)   HAWAII I.

Equator

MARQUESAS IS.

Ellice Ils.
Union Group

NAVIGATORS IS
Savaii   Upolu
Tutuila

LOW ARCHIPELAGO
Palliser Is.
SOCIETY ISLES   K. George's Is.
Tahiti   Resolution I.
COOKS ISLES   Dow I.

Tonga Is
Tongatabu

Tropic of Capricorn

Osnaburgh I.

Gambier Is.

Pitcairn I.

OCEANIA

AND

AUSTRALIA.

Chatham I.

crust which always surrounds their tissue. Thousands of islands in the ocean are of madreporic formation. The illustration represents half the natural size of a madrepore.

MADREPORE OF THE INDIAN OCEAN.

8. The Sandwich Islands are the only group in Polynesia that have attained to the dignity of nationality.

We have a treaty of friendship, commerce, and navigation with the king of the Sandwich Islands, who is a highly accomplished gentleman.

Honolulu, the capital (population 14,000), is the most important island town in the whole of Polynesia. It has an extensive trade with the United States.

Hawaii, the largest island of the group, is volcanic. On it towers, at the height of 13,760 feet, Mauna Loa, with its seething cauldron of molten lava, 1500 feet deep and two miles in circumference.

These islands are a famous place of rendezvous for our whalers.

Lot's wife, a shaft of granite rising 300 feet out of mid-ocean, is one of the wonders of Polynesia.

Formosa belongs to China. The French have occupied New Caledonia, the Society, and the Marquesas islands.

LOT'S WIFE.

Islands of the Atlantic Ocean. (Map, p. 152.)

9. All the islands in this ocean have, under the influences and agencies of commerce, been brought within the pale of civilization.

St. Helena, a rock, owned and fortified by Great Britain, is of note chiefly because it was there that Napoleon Bonaparte was imprisoned and died.

The Cape de Verdes, Madeira, and Azores all belong to Portugal. They are volcanic, and are famous for fruits and wines.

They produce also coffee, sugar, and tobacco. More than half the oranges imported into Great Britain are said to come from St. Michael's, one of the Azores. As many as 26,000 oranges have been known to be gathered from a single tree there in one year.

The Canaries, also volcanic, belong to Spain; they too are famous for their wines.

To them we owe the sweet little singing-bird that bears their name.

The Bermudas, said to consist of nearly 3,000 islets, belong to Great Britain. On them she has a naval station and an excellent dockyard.

LESSON LXXIX.

Study on the Map of Oceania and Australia.

Boundaries and Positions.—Between what parallels of latitude and meridians of longitude does Australia lie?—Bound New South Wales—Bound Victoria—North Australia—South Australia.—Bound West Australia. Where is Queen's-Land? Tasman Land?—Where is Tasmania?—New Zealand? What are their bearings from Melbourne? What tropic crosses Australia?—Where is Auckland I.?—Name the New Zealand Islands.

Point out the Sandwich Islands—the Low Archipelago—Fejee Islands—Gilbert Islands—New Hebrides—Admiralty Islands—Timor Islands—The Moluccas—The Ladrones—The Caroline Islands—Formosa—The Loo Choo Islands—Where are the Louisiade Islands?—New Caledonia?—The Solomon Isles?—Borneo?—Sumatra?—Java?—What groups form Micronesia?

SULTAN OF BORNEO

Seas, Gulfs, and Bays.—Where is the China Sea? *This is famous for the fierce typhoons that visit it.*—The Java Sea?—Celebes Sea?—Timor Sea?—Coral Sea?

Where is the Gulf of Australia?—Gulf of Carpentaria?—Spencer's Gulf?—
Temple Bay?—Princess Charlotte's Bay?—Shark Bay?—Halifax Bay?—Botany Bay?—Géographic Bay?—Storm Bay?—Jervis Bay?—Bay of Plenty?

Straits, Capes, Headlands, and Points.—Where is Sunda
Strait?—What does it separate?—Tell where the following straits are, and
what they separate:—Torres Strait—Dampier Strait—Malacca Strait—Strait
of Macassar—Bank Strait—Bass Strait—Foveaux Strait.

Where is Cape Leeuwin?—Cape Naturaliste?—West Cape?—Cape Bougainville?—Cape York?—Cape Coffin?—Cape Moreton?—Cape Melville?

Where is Java Head?—Macquarie Head?—Point Danger?—Rocky Point?

Mountains and Volcanoes, Rivers and Lakes.—Where are
the Australian Alps?—Where is Murchison Range?
—Mount Wilson?—Mount Egerton?—Mount Alexander?—Mount Cook?

Where is Mauna Loa? *Some of the lava rivers
ejected from Mauna Loa are twenty-six miles long.*
—Where is the Murray River?—The River Darling?—The Molyneux River?—Where is Lake
Eyre?—Lake Lefroy?

Cities and Distances.—Where is Melbourne?—Sidney?—Brisbane?—Adelaide?—
Hobart Town?—Auckland?—Batavia?—Borneo?—Manilla? *From Manilla we get Oranges.*—
Honolulu? *This is the chief entrepôt between the
opposite shores of the North Pacific.*—When it is 12 o'clock M. at Washington,
what is the hour at Peking?—At Sydney, Australia?—At San Francisco?

THE ORANGE.

TIME DIAGRAM.

GEOGRAPHICAL PROBLEMS.—1. A telegram is sent west from Calcutta to St. Louis. It is dispatched at
6 A. M. on the morning of January 1st, 1871, and passing over wires and cables is delivered in 1 hour and 6
minutes afterward in St. Louis: can you tell in what year and on what day and hour it reached St. Louis?
(Ans. See Time Diagram.)

II. Three friends part from each other in New York on the 1st day of January, 1870. A sails east to go
round the world; B journeys west, also to go round the world; and C remains in New York. On the
evening of December 31st, 1870, they meet again in New York, A and B having just completed the circuit
of the world. How many days has each seen in the year? Ans. A, 366; B, 364; and C, 365

LESSON LXXX.

The most Recent Geographical Events and Discoveries.

The record of geographical discoveries within the last few years is one of
great importance, and should be studiously pondered by the geographer.

(1.) The completion of the Pacific Railroad, by which the Atlantic and Pacific Oceans
were connected, was a geographical event which will always mark the year 1869.*

(2.) Within the last two years a new and extensive gold-district has been discovered
and examined in the vicinity of San Diego, California.

(3.) The Rio Colorado, the stream which drains the western slopes of the Rocky Mountains in Colorado and discharges its waters into the Pacific, has been explored to its highest
sources; the exploration revealing the ruins of cities, aqueducts, and fortified places belonging to a people of whose remote history we are ignorant.

(4.) The head-waters of the Missouri river have been traced into a region of great mineral wealth in Montana, and within less than a mile of a small stream which, by a devious
way, flows westward as a tributary of the Columbia.

(5.) The Physical Survey of several States of the Union has been undertaken and prosecuted with success. That conducted in Virginia has shown the great advantages to be derived from direct steam communication between her ports and European marts; her facilities for conveying, by canal and water-carriage, and consequently, at least expense (along
the James and Kanawha rivers to the Ohio, the Mississippi, and the upper Missouri), the
trade and emigration of the Old World, to the most distant parts of the West; and has
made known the physical geography and the unsurpassed mineral treasures of the State.

(6.) Renewed surveys of the Isthmus of Darien are in progress with a view to the construction, at no remote period, of a ship-canal to unite the Atlantic and the Pacific Oceans.

(7.) Observations made by officers of the English Cunard steamships show that the current which flows over the Banks of Newfoundland, has a mean temperature of 39° 2′ Fahr.,
during the three winter months falling to 32°, 31°, and 30°, and rising in September to 52°.

(8.) Deep-sea dredgings, by English geologists, in the Atlantic, reveal the astonishing
fact that animals supposed to have been extinct for ages are found living at great depths of
the ocean, and are busily engaged beneath the waters in the production of chalk, limestone,
and other rocks, and this occurs at depths once believed to be destitute of life.

(9.) About a year ago, an expedition, under Sir Samuel Baker, set out for the further exploration of Lake Albert Nyanza, in Africa, discovered by him in 1864.

(10.) In the fall of 1869, a communication from Dr. Livingstone, the celebrated African
traveller, announced the probable discovery of the long-sought sources of the Nile on elevated
lands southwest of Lake Tanganyika, in south latitude, between 10° and 12°, near the place
indicated by Ptolemy more than sixteen centuries ago. Should this be confirmed, the length
of the Nile, measured from a straight line from source to mouth, will be more than twice as
great as that of the Mississippi.

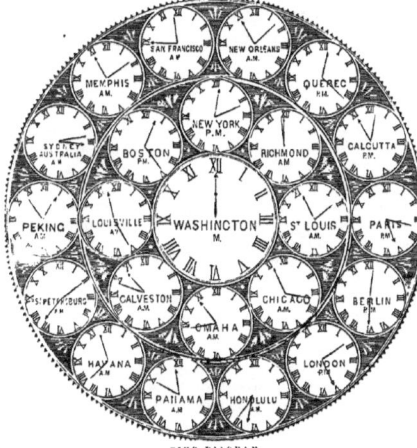

(11.) Rich diamond-mines have been discovered in Africa 800 miles from the Cape of
Good Hope.

(12.) Recent researches in Australia have proved the existence of extensive salt lakes,
soda deposits, and trees of enormous size. Immense beds of coal have been found in New
Zealand, which will go far toward shaping its industrial future.

(13.) In 1864, a Swedish expedition was sent by the government into the Arctic basin to
measure an arc of the meridian. It visited the southern shores of Spitzbergen, and, while
cruising around that island, confirmed the existence of Gillis Land, first seen, in 1707, by a
Dutch sea captain, whose name it bears. (See Gillis Land on Mercator's Map of the World.)

(14.) In 1868, a German expedition, fitted out at private expense, for the discovery of the
Pole, sailed in the steamer Germania, and reached the high point of 81° 5′ N. lat., and 10° E.
long. From the deck of the Germania the officers distinctly described, with the telescope,
the peaks of Gillis Land. They also found piles of drift-wood twenty feet high on the shores
of Spitzbergen, east up there by the currents of the ocean.

(15.) In the same year, another Swedish expedition penetrated as far as 81° 42′ N. lat.,
and brought back specimens of animal life at a depth of more than twenty-five hundred
fathoms, taken northwest of Spitzbergen. This expedition found drift-wood and vegetable
productions of the West Indies on the western and northwestern coasts of Spitzbergen.

(16.) In 1869, Captain Hall of Cincinnati returned from an Arctic voyage, bringing relics
and information of Sir John Franklin's long-lost and ill-fated expedition. Congress has made
an appropriation to send him out again on an expedition for Polar discoveries.

(17.) In the fall of 1870, an Hawaiian island reached San Francisco, with a cargo of 11,500
seal-skins, obtained on an uninhabited island in the waters adjacent to Alaska.

(18.) On the 11th of September, 1870, the Arctic Expedition in the Germania, before
alluded to, returned home. But they persevered, and in 1871 were rewarded with a good
view of "THE OPEN SEA in the Arctic Ocean."

NOTE.—It is of the utmost importance to geographical science that further explorations
be vigorously prosecuted for discoveries in the Antarctic circle. Within this circle is an area
equal to one-sixth the entire land-surface of our planet. It will be a reproach to the civilized
world to permit such a portion of the earth's surface any longer to remain Terra Incognita.

* The idea of connecting the two great oceans was first suggested by the Rev. James
Maury of Virginia, in a letter dated Louisa County, Jan. 10, 1736. In this letter he says:
"When it is considered how far the eastern branches of the Mississippi extend eastward,
and how near they come to the navigable rivers which empty themselves into the sea that
washes our shores to the east, it seems highly probable that its western branches reach as
far the other way, and make as near approaches to rivers emptying into the Pacific, across
a short and easy communication, opens itself to the navigator from that shore of the
continent unto the Eastern Indies."

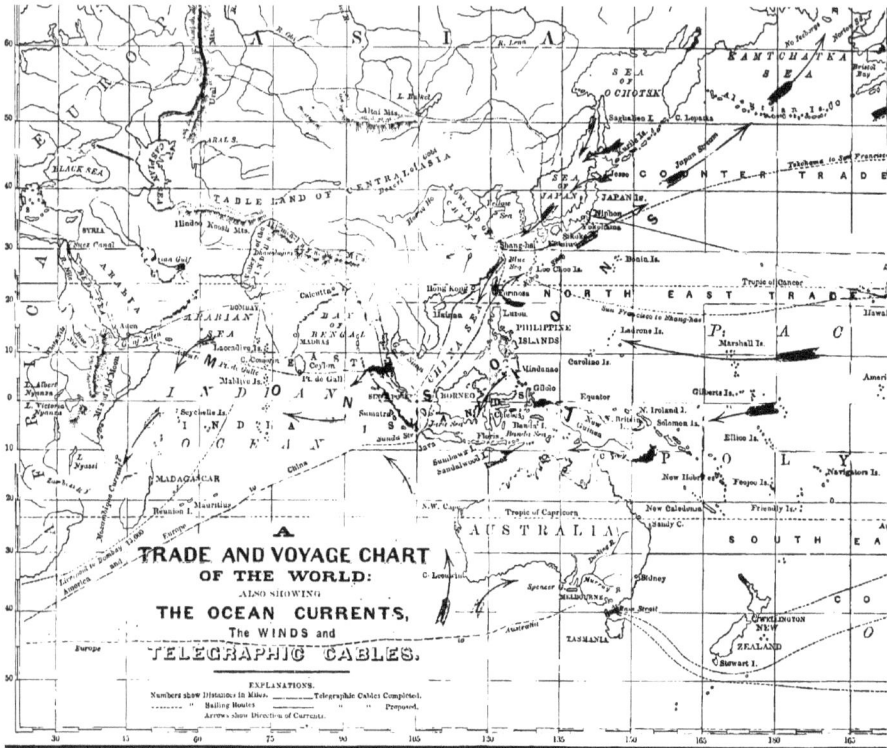

A

TRADE AND VOYAGE CHART
OF THE WORLD:
ALSO SHOWING

THE OCEAN CURRENTS,
The WINDS and
TELEGRAPHIC CABLES.

EXPLANATIONS.
Numbers show Distances in Miles. ———— Telegraphic Cables Completed.
·········· Sailing Routes ———— Proposed.
" Arrows show Direction of Currents.

The currents of the sea and the prevailing winds of the earth have already been alluded to. Upon a knowledge of these phenomena depends in a large measure the prosperity of all commercial nations. (See diagrams, pp. 110, 132.)

To avail himself of these winds and currents when fair, or to avoid them when adverse, is the effort of the master of every *sailing* vessel. With your present geographical information you can understand why the routes projected on the map are the routes usually chosen.

Questions.—Why do not navigators always sail in a direct course? What is the general direction of the winds in the northern half of the Torrid Zone? What in the southern half? What in the North Temperate Zone? In the South Temperate Zone? Is there any exception to this system? Point out the Monsoon regions of India. (See p. 132.)

The constant and permanent currents of the sea run between places where the waters are warm and places where the waters are cold, and for the reason

that the waters of the sea are, because cold in some places, warm in others, in constant state of unstable equilibrium. The warm Mozambique, the Gulf Stream and Japan Current, with the cold Humboldt and other Polar Currents, may be regarded as the unceasing effort of nature to restore the equilibrium which is i constant disturbance by the unequal distribution of heat over the ocean.

The left-hand edge, both of the Gulf Stream and Japan Current, is farthest t the north in autumn, farthest to the south in spring. In what zone would yo sail going from San Francisco to Hong Kong? In what zone, going from Shang hai to San Francisco?

The best route from New York to Liverpool is with the Gulf Stream, which helps vessels along at the average rate of twenty or thirty miles a day. Hov would the vessel make the quickest return passage? *Ans.* By keeping *out* of th Gulf Stream, going north of it in the fall, when there are no icebergs, or south o it in spring.

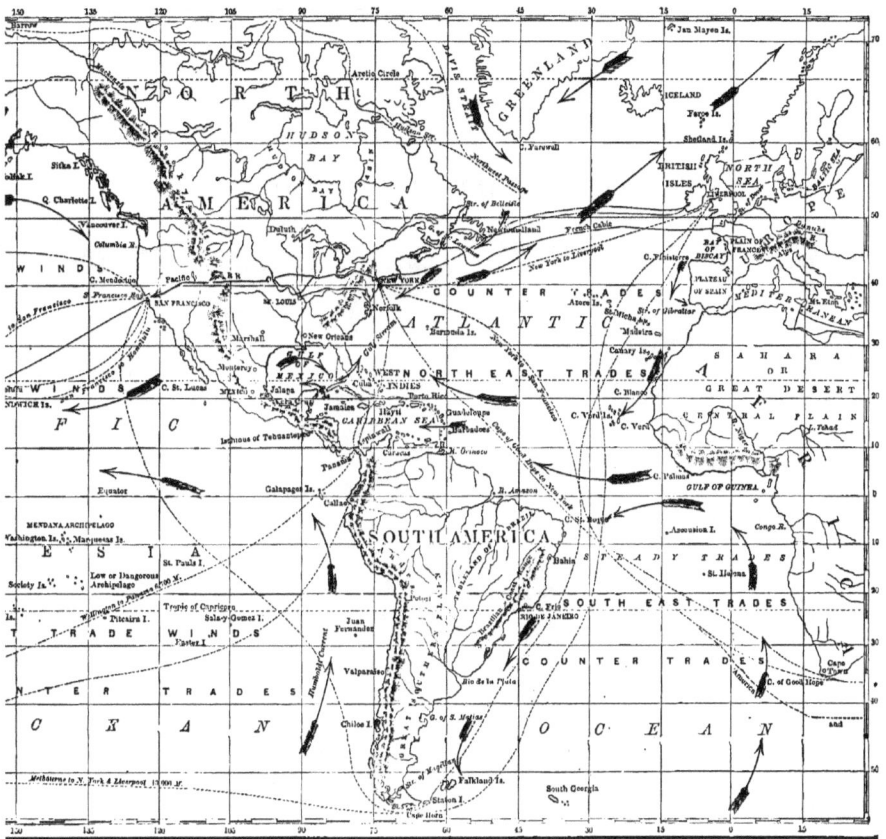

Trace the course of a vessel from New York to Aspinwall. From New York) San Francisco. Why do vessels bound from New York to San Francisco, after rossing the Equator in the Atlantic, stand in toward the coast of South America? Thy, after crossing the Equator in the Pacific, do they steer so far west? *Ans.* ecause they are forced by the northeast Trade-Winds. What winds bring them :to San Francisco? *Ans.* The Counter-Trades.

How would you go from San Francisco to the Sandwich Isles? What winds ould assist you soon after leaving port? *Ans.* The northeast Trades. How ould you return from these Islands to San Francisco? Why do you go north? *ns.* To escape the Trades that are now head-winds, and to catch the Counter-rades. How do you go from New York to Melbourne? *Ans.* By the Cape of ood Hope. Why would you remain on the eastern side of the Atlantic until m cross the Tropic of Capricorn? *Ans.* Because the southeast Trade-Winds mpel you.

In returning from Melbourne to New York, would you go by way of the Cape of Good Hope or Cape Horn? Why by Cape Horn? *Ans.* Because thus you have the Counter-Trades in your favor. Why do vessels from England bound to India, China, and Australia go by the Cape of Good Hope? *Ans.* Because the Counter-Trades assist them, and if they went by Cape Horn the Counter-Trades would be against them.

Trace the course of a vessel going from Liverpool to India. Why does she run across the Atlantic so close to the shores of South America? *Ans.* Because the Equatorial Current and the Trade-Winds force her over.

What is the shortest route from Liverpool to Bombay? *Ans.* By the Suez Canal. Through what waters would you take this route? How would you go from New York to San Francisco by rail? Point out the chief Ocean Telegraphic Cables of the world. Point out the long-sought Northwest Passage. Capt. McClure, of the British Navy, is the only explorer who has made this passage.

PRONOUNCING VOCABULARY.

In this Vocabulary the best and most recent authorities have been consulted for both spelling and pronunciation. ā, ē, ī, ō, ū, are to be pronounced as in bate, mete, bite, note, tube; ă, ĕ, ĭ, ŏ, ŭ, as in băt, bet, bit, not, but. The sound of a in far is indicated by ah; a in fall, by aw; o in do, by oo; g in get, by gh. The nasal sound occurring in some French words is indicated by N, as Toulon, (too-loN'); this nasal sound is somewhat like that of ng sounded through the nose. Letters enclosed by ( ) indicate pronunciation.

A.

Aalborg (ol'borg).
Aar (ahr).
Aarhuus (ar'hoos).
Abyssinia (ab-is-sin'i-a).
Aberdeen'.
Abomey (ah-o-mā').
Acapulco (ah-kah-pool'ko).
Accra (Italy) (ah-cher'rah).
Aconcagua (ah-kon-kah'gwa).
Aden (ā'den or ah'den).
Adige (ad'e-je ; It. ah'de-ja).
Ad-i-ron'dack.
Ad-ri-an-o'ple.
Ad-ri-at'ic.
Ægean (ē-jē'an).
Afghanistan (af-ghah-nis-tahn').
Af-ri-ca.
Agulhas (a-gool'yas).
Aix-la-Chapelle (āks-lah-shah-pel').
Ajaccio (ah-yaht'cho).
Alabama (al-a-bah'ma).
Aland (ah'land).
A-las-ka.
Albans (awl'benz).
Al-ba-marle'.
Al-cī'ra (Sp. ahl-thē'rah).
Aleutian (al-oo'she-an).
A'chi-son.
Albuquerque (ahl-boo-ker'kā).
Al'der-ney.
Aleppo (ah-lep'son).
A-lep'po.
Algiers (al-jērz').
Al-ge'ria.
Al'l-gha'ny.
Allahabad (ahl-la-hah-bahd').
Almaden (ahl-mah-den').
Alsace (ahl-sas').
Altai (ahl-tī').
Altamaha (al-ta-ma-haw').
Altona (al'to-na).
Altmuhl (ahlt'mühl). . . .
Amarapura (ahm-ah-ra-poo'ra).
Am'boise (Fr. ahN'bwahz).
Am-boy'na.
Am'i-ens (Fr. prou. ah-me-aN N').
Am'a-zon.
Amoo (ah-moo').
Amoor (ah-moor').
Amoy (ah'moy).
Amoor River.
Anady (an-a-gāh'rah).
Anadyr (ang'gē-se).
Ani-ap'o-lis.
Ann Ar'bor.
Antigua (ahn-tē'gwah).
Ant-arc'tic.
An-til'les (or ahn-tēl').
Ap-pa-lach'ee.
Ap-pa-la-chi-co'la.
Ap'en-nines.
A-ra'bi-a.
Ar'al.
Ar-a-rat.
Ar-au-ca'nia.
Archangel (ark-ān'gwi').
Archangel (ark-ān'jel).
Archipelago (ark-i-pel'a-go).
Arctic.
Ardennes (ar-dēn').
Arequipa (ah-ra-kee'pah).
Argentine (ar'jen-tēn).
Ar-kan'sas.
Armagh (ar'mah or ar-mah').
Arnentier (ar'jen-tēr).
As-cen'sion.
Ash-an-tee.
Asia (a'shi-ah).

As'eam (or as-sam').
As-sump'tion.
As-tra-khan'.
Atacama (ah-tah-kah'mah).
Atchafalaya (ach-af-a-lī'a).
Ath-a-bas'ca.
At-lan'tic.
Augustine (au-gus-tēn').
Aus-tra'li-a.
Aus-tm'ri-a.
Aus'tri-a.
Auvergne (o-vairn').
Aurungabad (o-rung-ga-bahd').
Avignon (ah-vēn-yoN').
Ava (ah'vah).
Az'ov.
Az-ores'.

B.

Balboa (bahl'boe).
Bal-el-Man'deb.
Baden (bah'den or bad'en).
Ba-ha'ma.
Bahia Honda (baia-ā' ah ōn'da).
Bahreïn (bah-rān').
Baïkal (bī'kahl).
Balize (bah'ka-kan').
Balkan (bah'kahn').
Balmoral (bah'mo-rahl).
Baltic (baw'tic).
Balzac (bahl-zahk').
Banbiaca (bah-bih-kiah'vah).
Bang-kok' or Ban-kok'.
Barataria (bah-ra-tah'rī-a).
Bar-ba'does (-dōz).
Barcelona (bar-se-lo'nah).
Barnaul (bar-nowl').
Ba-ta'vi-a.
Baton Rouge (bat'un roozh).
Ba-va'ri-a.
Bayonne (bah-yon').
Bayou la Pourche (bī'oo lah fourch).
Beaufort (bu'fort).
Beirut (bā'root).
Beled-el-jerid (bel'ed-el-jer'-eed).
Belém (be-lem' or bā-leN').
Belgi-um.
Bellelsle (bel-īle').
Bel-oo-chis-tan' (-tahn).
Bel-grade'.
Benares (ben-ah'rēz).
Bengal (beng-gawl').
Benguela (ben-gā'lah).
Benin (ben-ēn').
Ber'gen.
Ber-mu'das.
Berne (bern).
Behring (bēr'ring).
Ber-mud'.
Besançon (bā-sahn-soN').
Biafra (bē-af'rah).
Biél'de-ford.
Binghamton (bing'am-ton).
Bir'mah.
Biscay (bis'ka').
Bienh-tm (blen'im).
Bogota (bo-go-tah').
Bo-he'mi-a.
Boise (bois').
Bokhara (bo-kah'ra).
Bo-li'vi-a.
Bom-bay'.
Bonifacio (bon-e-fah'cho).
Bonin (bo-nēn').
Bordeaux (bor-do').
Bor'ne-o.
Born'holm.
Bos'po-rus.
Both'ni-a.
Bourbon (hoor'bon).
Bosna Serai (boz'nah ser-ī').
Bourges (boorzh).
Brah-ma-poo'tm.
Brazil (bra-zil').
Brazos (brah'zōs).
Bremen (in U. S. Brē'men).
Bre'men.
Bres'lau.
Brescia (bresh'e-ah).
Brisbane (briz'bane).

C.

Cabool (kah-hool').
Ca'diz.
Ca'en (or kahN).
Cal-fra'ri-a.
Ca-hau'ba.
Cairo (kī'ro ; in U. S, kā'ro).
Cithais (kal'īs ; Fr. prou. cal-ay').
Calcaśieu (kahl'ka-shoo).
Caldera (kal-dā'rah).
Cal-i-for'ni-a.
Callao (kahl-lah'ō, or kahl-yah'ō).
Cam'broy.
Cam'bri-a.
Cam-bo'di-a.
Can-ar'.
Ca-na'ver-al.
Can-ta'bri-a.
Can-ton' (in U. S. can'ton).
Cape Bre'ton (or brit'tn).
Caracas (cah-rah'cas).
Car-pen-ta'ri-a.
Car-ib-be'an.
Car-ta-ge'na.
Cas'pi-an.
Cas-siquiare (cah-see-kee-ah'rā).
Cat-te-ga.
Caxamarca (cah-hah-mar'kah).
Cayenne (kī-en').
Celebes (sel'e-bes).
Ceph-a-lo'ni-a.
Ceuta (sa'tah).
Ceylon (-ē'lon, or se-lon').
Cevennes (-sā-ven').
Chagres (chah'gres).
Chartres (shartr).
Champlain (sham-plane').
Charybdis (ka-rib'dis).
Chat-ta-hoo'chee.
Chaudière (sho-de-air').
Chelsea (chel'se).
Chenango (she-nang'go).
Cherbourg (sher-hoorg').
Cheyenne (she-en').
Cherapungee (cher-ah-poon-ga').
Ches'a-peake.
Chicago (che-caw'go).
Chili (chil'le).
Chimhorazo (chim-bo-rah'zo).
Chi'na.
Chir'cha.
Chihuahua (che-wah'wah).
Chili-li-ahh't.
Chine-e (chil-nēz').
Chip'pe-wa.
Christiania (kris-te-ah'ne-ah).
Chuquibamba (chu-ke-bahm'bah).
Chuquisaca (chu-ke-sah'kah).
Cienfuegos (se-en-fwā'gōs).
Cincinnati (sin-sin-nah'ti).
Cin'tra.
Cobija (ko-bē'hah).
Coahuila (ko-ah-wē'lah).
El Dorado (do-rah'do).
Cojutepeque (ko-hu-ta-pa'ka).
Colombia (ko-lom'be-a).
Comorin (kom'o-rin).
Com'cho.
Concement (con-mēnt'-ent).
Con-stan-ti-no'ple.
Concord (kong'kord).
Brennen (a-ta-ēn').
Copiapo (ko-pe-ah'po).
Co-pen-ha'gen.
Coquimbo (ko-kēm'bo).

Cur'do-va.
Corpus Christi (kris'tī).
Cor'si-ca.
Cot-ti-er'tes.
Coregoine (kew-s-gwe'nah).
Costa Rica (kos'tah re'kah).
Cotopaxi (co-to pax'i).
Cracow (krah'ko).
Cronstadt (krōn'-taht).
Covington (kov'ing-ton).
Crim'e-a.
Cu'ba.
Cuernes de Vera (kwa'vah dā va'rāh).
Cumana (ku-mah-nah').
Curacoa (ku-ra-co-a).
Cut-ti (kutch).
Cuzco (koos'ko).
Cycladce (sik'la-dēz).
Cy-clone'.
Cy'prus.

D.

Dah-lon'e-ga.
Dahu (dah-hoo').
Dakota (dah-ko'tah).
Dahomey (dah-ho'mā).
Dalton (dawl'ton).
Damine (dah'nel-dōz).
Dan'zic.
Dan'ube.
Dar-da-nelles'.
Darfur (dar-foor').
Da-ri-en'.
Dauphine (dō-fe'nā).
Dec'can.
Del'a-ware.
Delhi (del't'hī in U. S.; del c in A-idā)
Ben'mark.
Des Moines (dā moin').
De-troit'.
Dhwawalaghiri(dh-wol-a-ghe'-ree).
Diamantina (de-ah-mahn-te'nah).
Dieppe (dē-ēp').
Dijon (de-zhoN').
Dniéper (nē'per).
Dniester (nē'ter).
Do'fre-field.
Dominica (dom-e-nee'kah).
Dona Ana (do'nah ah'nah).
Dor'ches-ter.
Dorpat (dor'paht).
Douai (doo-ā').
Douro (du'ro).
Drontheim (dront'īm).
Druze (drūz).
Dungnacy (dung'aus-by).
Dun'kirk.
Dus'sel-dorf.
Dwina (dwe'nah).

E.

E'bro.
Ecuador (ek-wa-dōr').
Edgecombe (ej'kum).
Egypt (e'jipt or -burro).
Ed'is-to.
El'be.
Elburz (el-boorz').
El Dorado (do-rah'do).
El'sin-ore.
En'gland (īng'gland).
E-phe'sus.
Erze-rum.
Es'qui-mau (es'ke-mo).
Es-ma-ralda (es-ta-mah-dno'-rah).
Es-sequibo (es-se-kee'bo).
Es'tre-ma-du'ra.
Etienne (a-te-ēn').
Etna.
Eu-bœa (ū-bē'ah).
Eu-phra'tes.
Eure (ūr).
Eu-re'ka.

Eux'ine.
Ex'an-eville.
Ex'e-ter.
Ey're (air).

F.

Fænza (fah-ēn'zah).
Falkland (fawk'land).
Falmouth (fāl'muth).
Fa'roe.
Fayal (fī-awl'; Port. fī-ahl').
Fee'jee.
Felipe (fā-lē'pā).
Fernandina (-de'nah).
Ferrara (fer-rah'rah).
Ferrol (fer-rōl').
Fez'zan.
Fiesole (fē-es'o-lā).
Fiord (fī-ord').
Flint-terre (flin-te-tair').
Flo'ri-da.
Fond-du-Lac'.
For'mo'sa.
Foochoo (foo-choo').
Franche Comté (fraNsh koN-tā').
Frank'fort.
Fred'e-ricks-burg.
Freiburg (frī'boorg).
Fue-no (le foo-ā-hē'no).
Fuego (foo-ā'gō).
Fu'nen.
Furneaux (foor-no').

G.

Gal-a-qa'gos.
Galatz (gah-lahtz').
Galicia (ga-lish'e-a).
Galilee (gal'le'lee).
Galloway (gal-lo-wa').
Gambia (gam'bē-a).
Ganges (gan'jēz).
Garonne (gah-ron').
Gas-con'ade'.
Gas-co-ny.
Ge'ne-va.
Gen-o-a.
Geneseo (jen-e-se'o).
Gen'o-a.
Georg'e-a.
Gesul (zhā'raht).
Ger'ma-ny.
Ghauts (gauts).
Ghent.
Gi'bral-tar.
Gi'la (jē'la or hē'la).
Glou-ces'ter (glos'ter).
Gō-a.
Gob'i (gō'bē).
Godavery (go-dah'ver-ē).
Gon'dar (daz).
Gracias a Dios (grah'se-as ah dē'ōs).
Grafen-berg.
Grave-send.
Greenwich (grēn'ij or grēn'-itch).
Gren'a-da.
Greitz (grītz).
Grieshach (grēs'tiah).
Gri'qua-land.
Gua Ventre (grō'waNtr).
Guadalaxara (gwah-dah-lah-hah'rah).
Guadaloupe (gwa-dal-oop').
Guadalquiver (gaw-dal-quiv'-er).
Guadiana (gaw-de-ah'na).
Gualaspe (gwa'de-lupe).
Guadiana (gaw-de-ah'na).
Guam.
Guanajuato (gwah-nah-hwah't'o).
Guayaquil (gwi-ah-kē'l').
Guinea (gin'e).
Gu-ay-ra (gwī'rā).
Guiana (gā'ah-nah).
Gu'lf.
Gur-hwal (goor-dah-āl').
Guz-e-rat (goo-za-raht').
Gwalior (gwah'le-or').
Gwa-li-or.
Gyandotte (ghī-an-dot').

H.

Hack'en-sack.
Hadramaut (hahd-rah-mowt').
Hague (hāg).
Hainan (hī-nan').
Harz, or Hartz (harts).
Hakodadi (hah-ko-dah'di).
Ha'gers-town.
Hal'i-fax.
Ham'mer-fest.
Ham'burg.
Han'o-ver.
Harbor Grace (grahs).
Hat'ter-as.
Havana (ha-van nah).
Havre (hah'vr).
Haverhill (hav'er-il; in Eng., hāv'ril).
Hawaii (hah-wah'e).
Hayti (hā'tī).
Hebri-des (-dēz).
Hec'la.
Hel'go-land.
Helena (hel'e-na, Ark.; he-lē'na, for the island St. Helena).
Hel'sing-fors.
Hen-lo'pen.
Her-cu-la'ne-um.
Hermhut (hern'hoot).
Himalaya (hīm-a-lā'ya).
Hin-doo-koosh'.
Hin-doo-stan', or Hin-du-stan'.
Hoang-Ho (hwang-hō).
Hel'd'r-a-berg.
Hol'land.
Hol'stein (hōl'-tēn).
Holyoke (hōl'yōk).
Hon-du'ras.
Hong Kong'.
Honolulu (ho-no-loo'loo).
Hoo-sa-ton'ic (hoo-).
Houston (huls'tun).
Huahnga (hwah-āh'gah).
Huanuco (wah'noo-ko).
Hue (hoo'ā).
Hudra (hoof'resh).
Hum'boldt.
Hun'ga-ry.
Hu'ron.
Hyderabad (hī-der-a-bahd').

I.

Ice'land.
I'da.
Iguape (ē-gwah'pā).
Illinois (ill-in-oiz or -oi).
Il-la-wat'a, or Il-li-wat'a.
Inde-pen-dence.
In-di-an-ap'o-lis.
In-di-a-na.
Indre (ho-āah'na).
In'dus.
In'dri.
Inkerman (ink'er-man).
Interlachen (in'ter-lahk en).
In-ver-ness'.
I-o'wa.
Ire'land.
Irkutsk (ir-kootsk').
Ir'ra-wad-dy.
Isère (ē-zair').
Is'lip.
Is-pa-han'.
Is'tri-a.
Isthmus (is'mus).
It'a-ly.
Ivica (e've'cah).
Izmir (es-kōo'der).
Iztaccihuatl (ist-ak-see-hwah't'l).

J.

Jaen (hah-en').
Jalapa (hah-lah'pah).
Jal-an-cit'ra (yal-).
Jamaica (ja-ma'kah).
Jan Mayen (yahn mī'en).

Ja-pan'.
Japan (hah-poo'rah).
Jasey (ras'sl).
Java (jäh'vah).
Jeh'el sham'mer.
Je'na (or yā'nah).
Je-ru'sa-lem.
Jilton (he-lo'kol).
Joannes (zho-ahn'nes).
Jo'il-et.
Jornlic (ho-rool yo).
Ju-an Fer-nan'dez.
Ju'an de Fu'ca.
Juggernaut.
Junista (ju-nī'ah'tah).
Jungfrau (young'frow).
Ju'ra (Fr. zhu-rah').
Jut'land.

K.

Kafir (cah'ffr).
Kalahari (kah-lah-hah're).
Kul'a-ma-zoo.
Kam'-chat'ka.
Kanawha (kan-aw'wah).
Kan-da-har'.
Kan-ka-kee'.
Kan'sas (-zas).
Kari'stadt.
Kash'gar.
Ka-zan'Ki-a.
Ka-tah'din.
Kelat' (or -aht).
K-'nl-a.
Ken-tuck'y.
Ke'o-kuk.
Kerguelen (kerg'e-len).
Ke-wee'na-w.
Khar-toum'.
Khiva (kee'vah).
Kioukra (kē-ahk'tah).
Kiel (kēl).
Kiev (kē-ev').
Kilimandjaro (kil-e-mahn-
   jar'o).
King'ston.
Kirghis (kir-ghēz').
Kis'sin-gen.
Kin-lu (kee-oo'u-oo).
Kokan (ko-kahn').
Ko'mo-ro.
Kon'igs-burg.
Konka (kou'kah).
Kor-do-fan'.
Kron-stadt (krōn'staht).
Kuen Lun (kwen loon).
Kurdistan (koor-dis-tahn').
Kurile (koo-ril' or koo'ril).
Ku'ro-Si'wo (-si'wo).
Kur-ra-chee'.
Kwich'pak.

L.

Lniland (law'land, or lol'and).
Lab ra-dor'.
Lachen (lah'ken).
La-'co-dive.
La-do'ga.
La-drones'.
La Grange (-grǎnj).
La Guayra (lah gwi'rah).
Lahore (lah-hōr').
Lahn (lah-hoo').
La M orchs (-m shu'chah).
Landes (laNdz).
La Plat (lah plah'tah).
Lap'land.
Laramie (lar'a-me).
Lar'sa.
Lausanne (lo-zahn').
Leg'horn.
Leip'sic.
Lena (le'nah).
Lewes (lu'es).
Leyden (li'den or li'den).
Li-ber'ri-a.
Lichtenfels (lik'ten-fels).
Li-ege (lēj).
Lima (le'mah).
Limoges (le-mōzh').
Linyanti (lin-yahn'te).
Lipari (lip'a-re).
Liv'hon.
Liv'er-pool.
Llano Estacado (lyah'no e-
   tah-kah'do).
Lo-fo'e.
Lo-fo'den.
Loire (lwahr).
Lom'bar-dy.
Lo'mond.
Los Angeles (lōs an'jel-e-).
Lough (loh).
Louis-ville (loo'i-vil or loo'i-
   vil).
Lou-is'i-an-a (loo-'gl-ah'na).
Low'ell.
Lu'bec.
Luck'now.
Lu-pah'ta.
Luzon (loo-zōn' or -zon').
Ly'ons.

M.

Maas (mahs).
Mucns (ma-cah'n).

Ma-cas'sar.
Mack-cu'zie.
Mack-i-naw'.
Mad-a-gas'car.
Madeira (ma-de'ra).
Ma-dras'.
Mad-rid'.
Maestricht (mahs'trikt).
Mag-da-le'na.
Magalhaens (mah-gahl-yah'-
   ens).
Magellan (ma-jel'lan or maj-
   el-lan').
Maggiore (mahd-jo'rā).
Malmaitchin (mi-mi-chin').
Mal-a-bar'.
Ma-lac'ca.
Mal'a-ga.
Ma-lay'el-a.
Mal'dive.
Mamore (mah-mo'rā).
Ma-nil'la or Ma-nil'a.
Mant-choo-ri'a.
Marajo (mah-rah-zho').
Mar-an-ham'.
Margarita (mar-gah-re'tah).
Mar'mo-ra.
Marquesas (mar-ka'zas).
Mar-a-cay'bo.
Marseille (mar-sāl'y').
Mar-ta-han'.
Ma'ry-land.
Mas-a-chu'setts.
Mat-a-mo'ras.
Ma-tan'zas.
Mat'a-pan'.
Mauch Chunk (mawk chuugk)
Man'mee.
Mauritius (mau-rish'e-us).
Ma-zat'lan.
Mayenne (mah-yen').
Mechlin (mek'lin).
Med-i-ter-ra'ne-an.
Mec'ca.
Medina (me-de'nah in Ar., me-
   di'nah in U S.)
Melbourne (mel'burn).
Metuan (ma-uahm').
Me-nan'.
Men'ho (or mou'yo).
Metz (mets or māts).
Meuse (mŭza).
Merri-mack.
Messina (mes-se'nah).
Mex'i-co.
Miami (ml-an'I).
Michigan (mish'I-gan).
Mi'lan.
Mikado (ml-kah'do).
Mill'edge-ville.
Mindanao (min-dah-nah'o).
Mis-s-is-sip'pl.
Mis-sou'ri.
Mobile (mo-heel').
Mocha (mo'kah).
Mo-de'na.
Mo'hawk.
Mol-da'vi-a.
Mo-nee'a-cy.
Mont Cenis (cen'I).
Mo-lu'ca.
Mon-gō'li-a.
Mon-mo-ga-heu'lo.
Monterey (mont-e-rā').
Mont-e-vid'e-o.
Mont-gom'e-ry.
Montpelier (mont-pēl'yer).
Montreal (mon-tre-awl').
Mo-re'a.
Mo-roc'co.
Mosciuw (mos-ko').
Mozambique (mo-zam-bēk').
Munich (mu'nik).
Mur'ray.
Mus-cat'.

N.

Nac-og-do'ches.
Nagasaki (nah-gah-sah'ke).
Namaqua (nah-mah'qua).
Nan'ling.
Nan-king'.
Nantes (nants).
Nan-tuck'et.
Naples (na'plz).
Nar-ra-gan'sett.
Nar-bonne'.
Nash'ville.
Na-tal'.
Natch'ez.
Natch-i-toch'es.
Navarino (nah-vah-re'no).
Navarre (nah-vahr').
Ne-bras'ka.
Neit-gher'ry.
Nemours (ne-moor').
Ne-o'sho.
Nip'i-sing.
Ner-bud'dah.
Neuve (noo-a'ves).
Neufchatel (nu-hah-tel').
Neuse (ndz).
Nevada (ne-vah'dah).
Newark (nu'ark).
New-found'land (or nu'fund-
   land).
New Granada (grah-nah'dah).

New Or'le-ans.
New Zea'land.
Ngami (n'gah'me).
N'i-ag-a-ra.
Nicaragua (nik-a-rah'gwah).
Nic'o-bar.
Niemen (ne'men).
Niger (nī'jer).
Ning'po.
Niph-on'.
Nij'ni Nov'go-rod (nizh'nl).
Niemes (nēm).
Nor'wich (or nor'ich in U.S. ;
   nor'ij in Eng.)
Nor'way.
No'va Scot'ia.
No'va Zem'bla.
Nu'bi-a.
Nyanza (nī-ahn'zah).

O.

Oahu (wah'hoo).
Oaxaca (wah-hah'kah).
O'be.
Oceanis (o-she-a'nl-sh).
Ochotsk (o'kotsk).
Oc-mul'gee.
O-co'nee.
O'der.
O-des'sa.
Og'dons-burg.
O-gee'chee.
O-hi'o.
O-ke-cho'bee.
O-ke-fi-no'koc.
Old'en-burg.
Omaha (o-ma-haw').
Oneida (o-nī'dah).
Onondaga (on-hu-dah'gah).
Ou-ta-ri-o.
Ontonagon (on-to-nagh'on).
O-por'to.
Opelousas (op-e-loo'sas).
Orense (o-ren'sā).
Orizaba (o-re-zah'bah).
Or'e-gon.
O-ri-no'co.
Ork'ney.
Or'muz.
Or'te-gal.
O-sage'.
Osaka (o-sah'kah).
Os-ceo-la (os-e-o'la).
Os-tend'.
Os-we'go.
Ottawa (ot'ta-wah).
O-zark'.
Ox'ford.

P.

Pa-cif'ic.
Pais'ley.
Pa'lem-bang'.
Pa-ler'mo.
Pal'es-tine.
Palo Alto (pah'lo ahl'to).
Pam'li-co.
Pam'pas.
Pan-a-ma' (-nmā).
Papua (pap'u-ah).
Para (pah-rah').
Paraguay (pah-rah-gwā' or
   pah-rah-gwī').
Par-a-mar'i-bo.
Parana (pah-rah-nah').
Parime (pah-re'mā).
Par'is.
Par'ma.
Pas-ca-gou'la (-goo-).
Pas-sa'ic.
Pa-s-sa-nia-quod'dy.
Pat-a-gor'ni-a.
Peeos (pa'kōs).
Pe-dee'.
Pe-king'.
Peiling (pa-ling').
Pembina (pem'he-nah).
Penine (pen'ine).
Penn-syl-va'ni-a.
Pe-noh'scot.
Pen-sa-co'la.
Pe-o'ri-a.
Per'nam-bu'co.
Persia (per'shi-ah).
Perm (pe-rcr').
Petropaulowski (pa-tro-pau-
   lowsk'I).
Pic Anethou (pēk sh-na-too')
Piction (pik'too').
Pisa (pe'zah).
Piura (pl-oo'rah).
Placentia (pla-sen'shI-ah).
Placoenibo (plah-măn').
Plateau (plah-to').
Poitiers (poi-tērz').
Po-pay-an' (pah-pi-).
Po-po-cat-a-petl'.
Port au Prince (port o prins).
Por'to Bel'lo.
Porto Rico (por'to re'ko).
Portsmouth (ports'muth).
Po-to'mac.
Po-to-si'.
Potrol (po-to-se' or po-to'se).
Poughkeepsie (po-kip'e-o).

Prague (prāg).
Prairie du Chien (pra're du
   shēn).
Prosidio (pra-ze'de-o).
Provence (pro-vahN').
Prussia (prush'i-ah).
Pruth (prooth).
Puebla (pweb'lah).
Punta (poon'tah).
Puerto (poo'er-too).
Puerto Principe (pwer'to
   prin're-pa).
Pulaski (pu-las'ke).
Pyr'en-ees.

Q.

Quebec (kwe-bek').
Quentin (kwen'tin).
Queretaro (ka-ra'tah-ro).
Quilon (ke'lo-ah).
Quincy (kwin'cy).
Quito (ke'to).

R.

Racine (ra-sēn').
Rae-non'.
Raleigh (raw'lo).
Rang-oon'.
Rap-pa-han'nock.
Rap-id-au'.
Itar'i-tan.
Reading (red'ing).
Reims or Rheims (rēmz).
Reikiavik (rī'kī-a-vik).
Reuss (rōs).
Richelieu (re'she-loo or rē-h-
   loo).
Rideau.
Rideau (re-do')
Rio Janeiro (rē'o zhan-a'ro or
   zē'o jah-nā'ro).
Rhine (rīn).
Rhode Island (rōd i'laud).
Rhone (rōn).
Riga (re'gah).
Rio de la Plata (rē'o dā lah
   plah'tah).
Rio Grande (rī'o grand in
   Texas ; rē'o grahn'dā in S.
   America).
Rio Negro (rē'o nā'gro or
   nē'o jah-nā'ro).
Rochelle (ro-shel').
Roch'es-ter.
Ro-nan'o.
Rome (rōm).
Ros-et'ta (ro-zet'tah).
Rot'ter-dam.
Rou-en' (roo-en' or roo-oN').
Rumania (roo-me-hwah'ee).
Russia (rush'I-ah).
Rus-tschuk (roos-chook').

S.

Sabine (sah-bēn').
Sa'ble.
Saco (saw'ko).
Sag-ue-nay (sag-e-nā').
Sa'ghalien (sag-hal'I-en or sag-
   ha-lē'en).
Saguenay (sahg-e-nā').
Sahara (sah-hah'rah).
Sahalo (sah-lah'jo).
Salonica (sal-o-ne'kah).
Sa'lem.
St. Augustine (aw-gus-tēn').
St. Clair (klair).
St. He-lo'na.
St. Law'rence.
St. Louis (loo'is or loo'e).
St. Paul de Lo-an'do.
St. Pe'ters-burg.
St. Pierre (reint pēr or saN
   pe-air').
San Bias (blah).
San'dal-wood.
San Sal'va-dor.
San'ta Fe' (fā).
San-tee'.
Santiago (sahn-tl-ah'go).
Sar-a-to'ga
Sarawak (sah-rah-wahk').

Sar-din'i-a.
Sas-katch'a-wan.
Sa-van'nah.
Sault St. Marie (soo sēnt ma'-
   rī).
Save (sahv or sāv).
Savoy (sav'oy or sah-voy').
Sax'o-ny.
Scheldt (skelt).
Schuylkill (skool'kll).
Schoharie (sko-hǎr'e).
Scio (-l'o).
Scioto (si-o'to).
Scot'land.
Scutari (skoo'tah-re or skoo-
   tah're).
So-has'to-pol.
Seine (sān).
Sen'e-ca.
Senegal (sen'e-gawl or sen-e-
   gawl').
No-ne-gam'bi-a.
Sin-o'pe.
Shanghai (shang'hī).
Shen-an-do'ah.
Shet'land.
Si'am.
Si-be'ri-a.
Sic'i-ly.
Sierra Leone (se-er'rah le-o'-
   ne).
Sierra Madre (se-er'rah mah'-
   dra).
Sierra Morena (se-er'rah mo-
   rā'nah).
Si'non.
Simoda (sl-mo'dah).
Sing-a-pore'.
Smyr'na.
Sofimoa (so-le'mōa).
Soo-loo'.
South C'ar-o-li'na.
Sorata (so-rah'tah).
Stanovoi (stahn-no-voi').
Stannton (stăn'ton in Va ;
   stǎn'ton or stahn'ton in
   Eng.)
Stet'tin.
Sten'ben-ville.
Stock'holm.
Stutt'gard.
Suleiman (soo-lā-mahn').
Sungari (sun-gah're).
Surat (soo-rat').
Susquehanna (sus-kwe-hah'na).
Suwanee (su-wah'nee).
Swe'den.
Sumatra (soo-mah'trah)
Sus-que-han'nah.
Swe'den.
Switz'er-land.
Syd'ney.
Sy'ra.
Syracuse (sir'a-kūs).
Syr'i-a.

T.

Tabago (tah-bah'go).
Ta'gus.
Tahiti (tah-he'te).
Tah-le-quah'.
Tal-la-has'see.
Tal-la-poo'sa.
Tallulah (tǎl-u'la).
Tangs'co (tam-pe'ko).
Tananarivo (tah-nah-nah-re-
   voo').
Tausro (tah-nah'ro).
Tanganyika (tahn-gahn-ye'-
   kah).
Taujler (tahn-jēr').
Tapajos (tah-pah'zhōs).
Tarifa (tah-re'fah).
Ta-cau'to.
Tar'ta-ry.
Tas-ma'nl-a.
Taunton (tahn'ton).
Taur'us.
Tchad (chad).
Teheran (te-he-rahn').
Tehuantepec (ta-whan'ta-
   pek).
Tchinkebees (tchook'chēz).
Tau-ar-iffe'.
Tenasserim (te-nas'ser-im).
Ten-nes-see'.
Terre Haute (tor'e hōt).
Tex'as.
Thames (temz).
Thian'shan (te-ahn' shahn).
Thibet (tib'et).
Thoune (to'hah).
Tient-sing (te-ent-sēng').
Tierra del Fuego (teer'ra del
   fwa'go).
Tif'lis (or tiI-lēs').
Tip-pe-ca-noe'.
Tivoli (tiv'o-ly).
Ty'grls.
Tim-buc-too'tor (tim-buc'too).
Timor (ti-mōr').
Tobolsk (to-to-kah'kah).
To-bo-kk'.
Tocantins (to-kahn-tēnz).
To-le'do.
Tom-big'bee.

To-ron'to.
Tor-tu'gas.
Toulon (too-loN').
Toulouse (too-looz').
Tours (loor).
Tran-syl-va'ni-a.
Traf-al-gar' (or tra-fal'gar).
Travancore (trav-ahn-kōr').
Trieste (tri'est).
Trin-i-dad'.
Trip'o-li.
Troyes (trwah).
Truzillo (tru-hēl'yo).
Tucson (took'eon).
Tundras (toon'drahs).
Tu'nis.
Tu'rin (or tu-rin').
Turkestan (toor-kes-tahn').
Turk'ey.
Tus-ca-loo'sa.
Tus'ca-ny.
Tuxla (tooxt'lah).
Tus-cum-bia.
Tuz-cu'eo.

U.

Ueayle (oo-kī-ah'lā).
Uist (vvist).
Ulm (Ger. pron. oolm).
I'm-ba'goug.
Unp'qua.
Uni'ded States.
Uer'nar'is (oo-per'na-vik).
U'p-sah (up-sah'la).
U'ral.
U'ral or Ou'ral.
Uruguay (u'ra-gway or -gwī).
U'tah.
Utrecht (yoo'trockt).

V.

Valdai (vahl'dī).
Valladolid (val-ah-do-lid').
Valparaiso (val-pa-rī'so).
Vancouver (van-koo'ver).
Varinas (vah-re'nahs).
Vaud (vō).
Vendome (voN-dōm').
Venezuela (ven-e-zwe'la).
Ven'ice.
Vera Cruz (va'rah krooz).
Verona (ver-mh'ho).
Ver-mont'.
Versailles (ver-sālz').
Ve-su'vi-us.
Viatka (ve-aht'kah).
Vich (vīk).
Vicks'burg.
Vic-to'ri-a.
Vienna (ve-en'nah).
Vindhya (vind'yah).
Vis'tu-la.
Vis'ti-n.
Vol'ga.

W.

Wabash (wah'bash).
Wachusett (waw-chu'set).
Wahsatch (wah-sach').
Walhalla (wal-la'kl-ah).
War'saw.
Wash'ing ton.
Wachita (wash'e-taw).
Wa-ter-loo'.
We'ser.
West Indies (in'dīz).
Wheel'ing.
West'min-ster.
Wilkesbarre (wil-ke'bar).
Win'ni-peg.
Winnipiscogee (win-e-pe-
   ssk'ē).
Wis-con'-in.
Worcester (woos'ter).
Wur'tem-barg.
Wyandot (wī-an-dōt').
Wy-o'ming.

X.

Xarayes (hah-rī'es).
Xeres (hay-ree').
Xingn (shin-goo').

Y.

Yakutsk (yah-kootsk').
Yang-tse-Kiang (yahng'tse
   ke-ang').
Yapura (yah-po'rah).
Yazoo (yah-zoo').
Yedo.
Yeniel (yen-l-sa'l).
Yokohama (yo-ko-hah'mah).
Yu-ca-tan'.

Z.

Zacatecas (zah-a-tā'kas).
Zambezi (zam-bā'ze).
Zante' ville.
Zan-guc-bar'.
Zan-zi-bar'.
Zealand (ze'land).
Zurich (zu'rik).
Zuyder Zee (zī'der zee).

GEOGRAPHICAL STATISTICS.

Many of the statistics usually given at the close of works of this order have been carefully interwoven in the text itself. The subjoined Tables are compiled from the best authorities and the most recent data. For many of these the Author is indebted especially to the Census of the United States, 1870, to the Census of Great Britain and of the Dominion, 1871 (the latter, with their valuable information, kindly furnished by Mr. Alpheus Todd, Parliamentary Librarian), and to Behms' Geographische Jahrbuch.

☞ The Class should be exercised on these Tables by such questions as these : What is the area of the earth ? How much land ? How much water ? Population ? What Nation owns most land ? Which rules most people ? What is the population of the United States ? Railroads, length of Rivers, heights of Mountains, &c., &c.

DIMENSIONS OF THE EARTH.

Polar Diameter	7,899½ miles.
Equatorial Diameter	7,925½ "
Equatorial Circumference	24,899 "
Superficial Area	196,861,750 square miles.

HEIGHTS OF CHIEF MOUNTAINS.

(Pupil will find these statistics in the table at bottom of pp. 20 and 21.)

LENGTH OF CHIEF RIVERS.

(Pupil will find these at bottom of pp. 20 and 21. He will also refer for additional information to the appropriate place in the text.)

TABLE OF ENGLISH MILES TO A DEGREE OF LONGITUDE FOR EVERY FIFTH DEGREE OF LATITUDE.

(See p. 14.)

AREAS OF GRAND DIVISIONS, OCEANS, AND UNEXPLORED REGIONS.

	Square Miles.		Square Miles.
Europe, with islands	3,846,628	Unexplored Polar	
Asia	17,361,571	Regions	10,000,000
Africa	11,556,293	Arctic Ocean	3,513,792
North America	9,041,151	Atlantic Ocean	25,000,000
South America	6,957,371	Pacific Ocean	70,000,000
Australasia and Polynesia	3,425,302	Indian Ocean	30,000,000
Total land surface	52,168,028	Total earth's surface	196,861,750

TABLE OF THE EXTENT, ELEVATION, AND DEPTH OF THE GREAT AMERICAN LAKES.

	Length Miles.	Breadth, Miles.	Depth in Feet.	Elevation above sea in Feet.	Area in Sq. Miles.
Lake Superior	400	80	850	650	32,000
Green Bay	100	20	500	600	2,000
Lake Michigan	320	70	1000	600	22,400
" Huron	240	80	1050	600	20,400
" St. Clair	30	18	20	570	360
" Erie	240	38	150	565	9,600
" Ontario	180	35	650	234	6,300

The lakes contain 11,300 cubic miles of water, or more than one-half of all the fresh water on the globe.

AREAS OF OTHER PRINCIPAL LAKES.

	Square Miles.		Square Miles.
Caspian Sea	147,000	Lake Nicaragua	1,000
Sea of Aral	31,400	Lake Wener	2,124
Lake Baikal	15,000	Great Salt Lake	1,900
Great Slave Lake	12,000	Great Bear Lake	10,000
Lake Winnipeg	7,000	Lake Geneva	82
Lake Tidicaca	4,000		

POPULATION AND AREA OF THE EIGHT LARGEST ISLANDS.

	Population.	Square Miles.
Australia	200,000	2,915,000
Borneo	1,200,000	285,001
New Guinea	1,000,000	153,000
Madagascar	3,000,000	232,000
Sumatra	2,600,000	172,000
New Zealand	301,752	106,259
Niphon	30,000,000	95,000
Great Britain	26,000,000	87,000

The United States owns no islands except the Aleutian Islands, and those that are near our own shores, called littoral islands. Neither does China, nor Turkey, nor Russia, nor Brazil, own any except their littoral islands. The United States has possessions only on the American Continent. Great Britain, on the contrary, has possessions in the four quarters of the globe—her smallest possessions being in Europe ; these are now as follows :

	Inhabitants.	Square Miles.
In Europe	31,629,000	124,000
In Africa	1,400,000	500,000
In Asia	150,000,000	981,000
In Australia and Polynesia	2,000,000	3,058,000
In America	6,130,000	3,597,355
Total	193,659,000	8,277,355

THE SIX NATIONS THAT OWN MORE THAN HALF THE LAND AND GOVERN MORE THAN HALF THE PEOPLE IN THE WORLD, are—

	Inhabitants.	Square Miles.
China	480,000,000	4,605,000
Great Britain	193,000,000	8,340,000
Russia	77,000,000	7,860,000
United States	39,000,000	3,614,000
Brazil	13,000,000	3,230,000
Turkey	41,000,000	1,820,000
Total	843,000,000	29,587,000

AREAS AND POPULATIONS OF THE UNITED STATES.

(By Census of 1870.)

States and Territories.	Square Miles.	Aggregate Population.	Whites.	Colored.	Chinese, Japanese and Indians.
Maine	35,000	626,915	624,809	1,606	500
N. Hampshire	9,280	318,300	317,697	580	23
Vermont	10,212	330,551	329,613	924	14
Massachusetts	7,800	1,457,351	1,443,156	13,947	218
Rhode Island	1,306	217,353	212,219	4,980	154
Connecticut	4,750	537,454	527,549	9,668	237
New England States, aggregate	68,348	3,487,924	3,455,043	31,765	1,156
New York	47,000	4,382,759	4,330,210	52,081	468
New Jersey	8,320	906,096	875,407	30,658	31
Pennsylvania	46,000	3,521,951	3,456,609	65,294	18
Delaware	2,120	125,015	102,221	22,794	
Maryland	11,124	780,894	605,497	175,391	6
District of Columbia	60	131,700	88,278	43,404	18
Mid. States, aggregate	114,614	9,848,115	9,458,242	389,622	571
Virginia	38,352	1,225,163	712,089	512,841	233
North Carolina	50,704	1,071,361	678,470	391,650	1,241
So. Carolina	34,000	705,606	289,667	415,814	125
Georgia	58,000	1,184,109	638,926	545,142	41
Florida	59,268	187,748	96,057	91,689	2
Alabama	50,722	996,992	521,384	475,510	98
Mississippi	47,156	827,922	382,896	444,201	825
Louisiana	41,346	726,915	362,065	364,210	640
Texas	274,356	818,579	564,700	253,475	404
Arkansas	52,198	484,471	362,115	122,169	187
Tennessee	45,600	1,258,520	936,119	322,331	70
Indian Territory (est.)	68,991	12,000			
Sou. Mexico	121,201	91,874	90,393	174	1,307
Southern States, agg	943,801	9,501,260	5,031,880	3,509,204	5,175
West Virginia	23,000	442,014	424,033	17,980	1
Ohio	39,964	2,665,260	2,601,946	63,213	101
Kentucky	37,680	1,321,011	1,098,692	222,210	109
Indiana	33,809	1,680,637	1,655,837	24,560	240
Illinois	55,405	2,539,891	2,511,096	28,762	33
Michigan	56,451	1,184,059	1,167,282	11,849	4,928
Wisconsin	53,924	1,054,670	1,051,351	2,113	1,206
Missouri	65,350	1,721,295	1,603,146	118,071	78
Iowa	55,045	1,191,020	1,188,207	5,562	51
Minnesota	83,531	439,706	438,257	759	690
Kansas	81,318	364,399	346,377	17,108	914
Nebraska	75,995	122,993	122,117	789	87
Colorado	104,500	39,864	39,221	456	187
Montana	143,776	20,595	18,306	183	2,106
Dakota	150,932	14,181	12,887	94	1,200
Wyoming	97,883	9,118	8,726	183	989
Western States, agg	1,258,567	11,813,713	11,599,161	544,992	12,150
California*	188,982	560,247	499,424	4,272	56,531
Oregon	95,274	90,923	86,929	346	3,648
Nevada	112,090	42,491	38,959	357	3,175
Arizona	113,916	9,658	9,581	26	51
Utah	84,176	86,786	86,044	118	445
Idaho	86,294	14,999	10,618	60	4,321
Washington	69,994	23,955	22,195	207	1,553
Alaska (est.)	577,390	6,000	6,000		
Pacific States, agg	1,338,416	885,059	759,750	5,386	69,943
Total Aggregate	3,611,819	38,555,371	33,595,374	4,880,009	88,985

* California has 49,310 Chinese.
* Exclusive of Navy, Mariners at Sea, and Wild Indians, estimated at 367,727.

GROWTH OF THE UNITED STATES IN STATES AND POPULATION.

	Population.			Population.	
1775	2,803,000 Colonies, 13		1830	12,866,020 States	23
1790	3,929,827 States	13	1840	17,069,453 "	25
1800	5,305,937 "	16	1850	23,191,876 "	30
1810	7,239,814 "	17	1860	31,719,705 "	33
1820	9,638,191 "	20	1870	38,876,371 "	37

OTHER PARTS OF AMERICA.

	Date.	Population.	Sq. Mile.
Greenland (estimated)	1871	10,000	730,790
Newfoundland	"	146,536	40,300
Labrador (estimated)	"	5,000	74,840
Prince Edward's Island	"	94,021	2,150
The Dominion	"	3,576,377	3,817,045
Ontario	"	1,620,842	419,020
Quebec	"	1,191,565	121,010
Nova Scotia	"	387,800	18,660
New Brunswick	"	285,777	27,163
Manitoba	"	11,963	13,500
British Columbia (est.)	"	50,000	220,000
Northwest Territories (est.)	"	48,500	2,797,060
Mexico	1870	9,150,000	773,110
Cent. America (p. 87), est	1868	2,690,635	188,363
West Indies (est.)	1871	4,000,000	93,860
Cuba	"	1,400,000	45,880
Land San Domingo (wst.)	"	710,000	48,030
South America (est.)	"	20,000,000	6,957,371
Brazil	1872	10,000,000	3,230,300
The Guyanas	1865	216,000	197,845
Venezuela	"	2,200,000	368,425
United States of Colombia	"	2,900,000	357,158
Ecuador	"	1,300,000	219,978
Peru	"	2,500,000	500,091
Bolivia	"	1,987,000	536,752
Chili	1867	2,085,000	132,619
Argentine Confederation	"	1,801,000	886,801
Paraguay	"	1,337,000	136,348
Uruguay	"	250,000	66,613
Patagonia	"	30,000	376,332

EUROPE.

	Date.	Population.	Sq. Miles.
British Isles	1871	31,465,480	120,760
England	"	20,982,316	50,922
Wales	"	1,217,135	7,398
Scotland	"	3,358,613	30,686
Ireland	"	5,402,759	31,754
New German Empire	1864	40,577,744	213,060
Prussia	"	36,521,412	156,068
Bavaria	"	4,925,000	29,622
Norway	"	2,845,864	5,708
Hanover	"	1,923,492	14,854
Wurtemberg	"	1,748,328	7,544
Austria	"	35,504,547	240,350
Hungary	"	10,081,854	82,485
Turkey in Europe	1866	9,460,000	110,669
France	1867	1,096,810	18,347
Ionian Republic	"	251,712	1,006
San Marino	1870	8,500	21
Switzerland	1860	2,510,494	15,509
Andorra (est.)	1871	12,000	150
Italy (estimated)	"	25,901,940	114,386
Spain	"	16,392,625	195,502
Portugal	1863	3,987,861	36,403
Belgium	1865	4,984,451	11,374
Holland	"	3,552,605	13,664
Denmark	"	1,948,645	14,733
Sweden	1864	4,050,611	170,627
Norway	1865	1,701,478	123,290
Russia in Europe	1861	58,241,892	2,116,760
Finland	"	1,751,803	48,962
Iceland	1866	1,844,600	145,316
The Caucasus	1863	4,157,917	108,780

ASIA.

	Date.	Population.	Sq. Miles.
British India	1865	195,300,114	1,576,611
Ceylon	1865	2,019,738	21,703
Farther India	1867	20,709,915	752,072
Burmah (est.)	"	4,000,000	190,531
Siam	"	5,200,390	300,014
Anam	"	9,000,000	198,020
French Cochin China	"	959,116	1,664
East India Islands (est.)	1871	27,878,894	854,910
Japan (est.)	"	35,000,000	149,384

ASIA.—Continued.

	Date.	Population.	Sq. Miles.
Chinese Empire (est)	1871	480,000,000	4,605,105
Russia in Asia (Siberia)	1863	1,625,690	5,584,767
Afghanistan (est)	1871	4,000,000	434,321
Beloochistan (est)	"	2,000,000	165,825
Arabia (est)	"	4,000,000	1,026,080
Persia (est)	"	5,000,000	502,827
Turkey (est)	"	17,000,000	667,744

AFRICA.

	Date.	Population.	Sq. Miles.
Egypt	1871	7,465,000	659,080
Tripoli	"	750,000	298,426
Tunis	"	950,000	7,640
Algeria	1867	2,929,216	258,809
Morocco (est)	1871	2,750,000	259,584
Abyssinia	1868	3,000,000	158,387
Liberia	1867	717,500	9,567
Cape Colony	"	496,381	192,750
Natal	"	156,165	19,317
Diamond Fields and the two Republics (est)	1871	500,000	300,000
Madagascar	1865	5,000,000	242,368

AUSTRALIA AND NEW ZEALAND.

	Date.	Population.	Sq. Miles.
Australian Colonies	1866	1,690,850	3,017,591
New South Wales	"	413,886	308,560
Victoria	"	626,870	88,452
South Australia	"	167,814	380,602
Queensland	"	87,775	668,250
West Australia	"	20,380	978,824
Tasmania	"	93,391	26,215
New Zealand	"	201,712	106,259

PRINCIPAL TOWNS AND CITIES OF THE UNITED STATES OF AMERICA.

(CENSUS OF 1870.)

Albany	N. Y.	69,422
Alexandria	Va.	13,570
Alleghany	Pa.	53,180
Atlanta	Ga.	21,789
Auburn	N. Y.	17,225
Augusta	Ga.	15,389
Austin	Tex.	4,428
Baltimore	Md.	267,354
Bangor	Me.	18,289
Boston	Mass.	250,526
Bridgeport	Conn.	18,969
Brooklyn	N. Y.	396,099
Buffalo	N. Y.	117,714
Burlington	Iowa	14,980
Burlington	Vt.	14,387
Cambridge	Mass.	39,634
Camden	N. J.	20,045
Charleston	S. C.	48,956
Charlestown	Mass.	28,323
Charlotte	N. C.	4,473
Chattanooga	Tenn.	6,093
Chicago	Ill.	298,977
Cincinnati	O.	216,239
Cleveland	O.	92,829
Columbia	S. C.	9,298
Columbus	Ga.	7,401
Columbus	O.	31,474
Concord	N. H.	12,241
Covington	Ky.	24,505
Davenport	Iowa	20,038
Dayton	O.	30,473
Des Moines	Iowa	12,035
Detroit	Mich.	79,577
Dubuque	Iowa	18,434
Elizabeth	N. J.	20,832
Elmira	N. Y.	15,863
Erie	Pa.	19,646
Evansville	Ind.	21,830
Fall River	Mass.	26,766
Fort Wayne	Ind.	17,718
Galveston	Tex.	13,818
Grand Rapids	Mich.	16,507
Hannibal	Mo.	10,143
Harrisburg	Pa.	23,104
Hartford	Conn.	37,180
Hoboken	N. J.	20,297
Houston	Tex.	9,382
Huntsville	Ala.	4,907
Indianapolis	Ind.	48,241
Jackson	Miss.	4,234
Jacksonville	Fla.	6,913
Jersey City	N. J.	82,546
Kansas City	Mo.	32,260
Knoxville	Tenn.	8,682
Lancaster	Pa.	20,233
Lawrence	Mass.	28,921
Leavenworth	Kan.	17,873
Lexington	Ky.	14,801
Little Rock	Ark.	12,380
Louisville	Ky.	100,753
Lowell	Mass.	40,928
Lynn	Mass.	28,233
Lynchburg	Va.	6,825
Macon	Ga.	10,810

Manchester	N. H.	23,536
Memphis	Tenn.	40,226
Milwaukee	Wis.	71,440
Minneapolis	Minn.	13,066
Mobile	Ala.	32,034
Montgomery	Ala.	10,588
Nashville	Tenn.	25,865
Natchez	Miss.	9,057
New Albany	Ind.	15,396
New Bedford	Mass.	21,340
Newark	N. J.	105,059
Newburyport	Mass.	21,595
New Brunswick	N. J.	15,058
New Haven	Conn.	50,840
New Orleans	La.	191,418
Newport	Ky.	15,087
New York	N. Y.	942,292
Norfolk	Va.	19,229
Norwich	Conn.	16,653
Omaha	Neb.	16,083
Oswego	N. Y.	20,910
Paterson	N. J.	33,579
Peoria	Ill.	22,849
Petersburg	Va.	18,952
Philadelphia	Pa.	674,022
Pittsburg	Pa.	86,076
Portland	Me.	31,413
Portland	Or.	8,293
Portsmouth	Va.	10,492
Poughkeepsie	N. Y.	20,081
Providence	R. I.	68,904
Quincy	Ill.	24,052
Raleigh	N. C.	7,790
Reading	Pa.	33,930
Richmond	Va.	51,038
Rochester	N. Y.	62,386
Sacramento	Cal.	16,283
Salt Lake City	Utah	12,851
St. Joseph	Mo.	19,565
St. Louis	Mo.	310,861
St. Paul	Minn.	20,030
Salem	Mass.	24,117
San Antonio	Tex.	12,256
San Francisco	Cal.	149,173
Savannah	Ga.	28,235
Scranton	Pa.	35,092
Selma	Ala.	6,484
Springfield	Ill.	17,364
Springfield	Mass.	26,703
Stockton	Cal.	10,066
Syracuse	N. Y.	43,051
Taunton	Mass.	18,629
Terre Haute	Ind.	16,103
Toledo	O.	31,584
Trenton	N. J.	22,874
Troy	N. Y.	46,465
Utica	N. Y.	28,804
Vicksburg	Miss.	12,443
Washington	D. C.	109,199
Wheeling	W. V.	19,280
Williamsport	Pa.	16,030
Wilmington	Del.	30,841
Worcester	Mass.	41,105

POPULATION OF SOME OF THE PRINCIPAL CITIES OF THE WORLD.

NORTH AMERICA.

[For chief Cities and Towns of the United States, see above.]

Montreal	Dom. of Canada	107,000
Quebec	"	60,000
Toronto	"	56,600
Ottawa	"	22,000
London	"	15,800
Halifax	"	27,000
Mexico	Mexico	210,000
Havana	Cuba	150,000

SOUTH AMERICA.

Bogota	U. S. of Colombia	45,000
Caracas	Venezuela	40,000
Quito	Ecuador	80,000
Lima	Peru	100,000
La Paz	Bolivia	75,000
Santiago	Chili	115,000
Valparaiso	"	80,000
Rio de Janeiro	Brazil	420,000
Bahia	"	180,000
Pernambuco	"	120,000
Buenos Ayres	Argentine Confed	200,000
Montevideo	Uruguay	100,000

EUROPE.

London	England	3,880,000
Liverpool	"	493,000
Manchester	"	355,000
Birmingham	"	343,000
Leeds	"	260,000
Sheffield	"	240,000
Bristol	"	182,000
Merthyr Tydvil	Wales	97,000
Glasgow	Scotland	477,000
Edinburg	"	196,500
Dublin	Ireland	246,000
Belfast	"	174,000
Cork	"	78,000
Paris	France	1,840,000
Lyons	"	323,000
Marseilles	"	300,000
Bordeaux	"	195,000
Nantes	"	155,000
Toulouse	"	115,000
Rouen	"	100,000
Havre	"	82,000

Berlin	German Empire	1,000,000
Hamburg	"	305,000
Munich	"	170,000
Breslau	"	170,000
Dresden	"	160,000
Cologne	"	126,000
Konigsberg	"	100,000
Leipsic	"	95,000
Bremen	"	75,000
Vienna	Austria	607,000
Prague	"	157,000
Trieste	"	105,000
Lemberg	"	79,000
Pesth	Hungary	157,000
Naples	Italy	420,000
Milan	"	250,000
Rome	"	220,000
Turin	"	204,000
Palermo	"	180,000
Genoa	"	130,000
Florence	"	120,000
Venice	"	120,000
Madrid	Spain	475,000
Barcelona	"	252,000
Seville	"	152,000
Valencia	"	145,000
Malaga	"	115,000
Granada	"	100,000
Cadiz	"	75,000
Lisbon	Portugal	295,000
Oporto	"	90,000
Brussels	Belgium	190,000
Ghent	"	125,000
Antwerp	"	125,000
Amsterdam	Holland	265,000
Rotterdam	"	120,000
Copenhagen	Denmark	160,000
Stockholm	Sweden	140,000
Christiana	Norway	60,000
St. Petersburg	Russia	550,000
Moscow	"	360,000
Warsaw	"	245,000
Odessa	"	120,000
Riga	"	100,000
Astrachan	"	45,000
Archangel	"	20,000
Geneva	Switzerland	40,000
Berne	"	30,000
Athens	Greece	44,000
Constantinople	Turkey	1,075,000
Adrianople	"	150,000
Salonica	"	90,000
Bucharest	"	125,000

ASIA.

Smyrna	Turkey	160,000
Damascus	"	150,000
Aleppo	"	100,000
Bagdad	"	65,000
Muscat	Arabia	60,000
Aden	"	50,000
Mecca	"	40,000
Teheran	Persia	100,000
Tabriz	"	100,000
Bokhara	Turkestan	125,000
Calcutta	India	1,000,000
Bombay	"	820,000
Madras	"	450,000
Benares	"	600,000
Lucknow	"	300,000
Delhi	"	160,000
Bankok	Siam	400,000
Singapore	"	70,000
Pekin	China	1,650,000
Canton	"	1,200,000
Shanghai	"	300,000
Nanking	"	800,000
Amoy	"	250,000
Yedo	Japan	1,550,000
Miako	"	370,000
Osaca	"	700,000
Yokohama	"	300,000

OCEANIA.

Manila	Philippine Islands	160,000
Batavia	Java	360,000
Melbourne	Australia	130,000
Sydney	"	100,000
Honolulu	Sandwich Islands	14,000

AFRICA.

Cairo	Egypt	260,000
Alexandria	"	175,000
Tunis	Tunis	150,000
Morocco	Morocco	100,000
Fez	"	75,000
Algiers	Algeria	60,000

ESTIMATED POPULATION OF THE EARTH IN 1871.

America	96,000,000	Australia, Poly-	
Europe	295,000,000	nesia and other	
Asia	800,000,000	Islands	5,000,000
Africa	190,000,000		
Total			1,386,000,000

MILES OF RAILROAD, 1871.

America	56,000 miles.	Africa	1,500 miles.
(United States	51,000 ")	Australia	1,000 "
Europe	61,000 "		
(Great Britain	16,000 ")	Total, World, 127,500 miles.	
Asia	5,000 "		

MAP-DRAWING.

I.—PRELIMINARY METHOD.

How to teach map-drawing is a perplexing question with many teachers. But systematic efforts, aided by a few simple rules, will soon make it a favorite exercise of the class-room.

Map Tracing.— Lay a transparent sheet (tracing paper) smoothly over the original. Secure it so that it may not slip, and then trace with pen or pencil what you wish to copy.

Map Sketching.—The object of this is to practice the eye and hand, as well as to impress upon the mind the geographical features of the country. It is one of the most useful exercises in the study of geography.

It is not expected that the pupil can, at first, draw from memory a map that will look just like the original in the book, but let each one draw something as much like it as he can. The best way is to begin with the original before him. Let us take an easy one, as North Carolina, and begin by examining its outlines.

Well, now look along the northern boundary, that is a straight line. Follow down the western boundary, which slants to the west. The southern line is broken; follow it carefully; and now measure with your pencil the width of the State along the coast. Now measure the width in the western part. Now measure the length, and see how much longer the State is than it is wide. Close your books. I shall give you *just five minutes* to draw this map; so you must work rapidly. James, John, Matthew, Henry, Charles, Robert, go to the blackboards, and the rest of you take your slates. Are you ready? Begin. *Draw the coast-line first,—sketch lightly and rapidly;*—then draw the northern boundary; next the western; and then the southern. In five minutes the signal is given. The work ceases. All are seated. Books are opened, and each map on the board is good-naturedly criticized. Now, scholars, I will draw the northern line of this map for you. I think it looks better, after you have traced it lightly, to take the blunt end of your chalk-crayon and go over it with *short heavy strokes,* thus : ━━

North Carolina

The next day.—Well, have you forgotten how North Carolina looks? No sir. Look at your geographies again. Can you draw the northern part of it for the length of an inch? Yes, sir. Can you then draw *another* inch? Yes, sir. Very well; so you can draw the whole of it inch by inch. Measure again the width of it with your pencil in the eastern part; now in the western part, and remember the difference. Now measure the length, and see how many times the width it measures. Now notice the mountains.

Next notice the beginning of the Neuse River, the directions from which and in which it flows, and where it empties.

Again, six boys at the blackboards and the rest with slates, have *five minutes* given to sketch the map. The boundaries are retraced according to yesterday's instruction.—Now, scholars, you may draw the rivers. Make a wavy or vibratory motion with your crayons, thus : ∼∼∼∼∼∼∼ Can you do this? A FEW MINUTES ARE SILENTLY SPENT IN PRACTICING.

Now, scholars, to-morrow North Carolina will be a special lesson, and I wish you to study the map, the rivers, location of the towns, etc., and impress upon your minds the features of the State, as you would those of a man you wished to remember for life.

[*Let pupils here refer to General Hints for Map-Drawing, p. 160.*]

North Carolina

The next day spend about ten minutes in drawing maps of North Carolina. Let the pupils represent mountains in the following manner, and after tracing lightly the coast-line, go over it with the blunt end of the crayon, making a broad, heavy stroke.

MOUNTAINS.

Now, scholars, to-morrow morning I want you all to bring in a map of North Carolina, drawn either with pen or pencil; and I prefer *that drawn with a pencil.*

ALWAYS REQUIRE MAPS TO BE DRAWN AS LARGE AS THE SLATE OR BLACKBOARD WILL ALLOW ; a better effect is produced and more freedom of the hand acquired ; the pupils' attention for the present being called particularly to the *comparative,* rather than to the absolute measurements. Let pupils use the book until they become experts, then require them to sketch from memory. They may indicate by a small cross the extreme N., S., E., W., N.E., N.W., S.E., and S.W. limits, measuring with their pencils, and comparing distances, and then connecting the various points with slight *dotted* lines, following as nearly as possible the contour of the map, afterward retracing with heavy lines. *Rapidity* rather than accuracy should be required at first; for *time is precious.* Accuracy will come with practice.

PUPILS MUST EXECUTE AT HOME EVERY MAP DRAWN IN THE CLASS-ROOM, BUT ONLY AFTER REPEATED DRAWINGS.

The knowledge of geography thus acquired, will have been attained through a process which trains the eye and the hand—giving judgment and skill, developing the perceptive faculties, and creating tastes of inestimable value to the individual.

MAP-DRAWING ON PAPER SHOULD ADVANCE, *pari passu,* WITH OFF-HAND MAP-DRAWING ON THE BLACKBOARD AND SLATE. After a map has been drawn and redrawn, and discussed in all its physical features before the class, it should be given as the special lesson of the next day, and every pupil be required to execute a map. The comparison of these maps will awaken a high degree of interest in the pupils. Gradually, as the work becomes familiar, details may be insisted upon. Require the paper to be of a certain size and form. Bristol-board is best ; and instruct the pupil to leave one and one-half inches of white margin, and to make marginal lines resembling those in their books.

MARGINAL LINES.

After some proficiency has been attained instruction may be given in making the wavy lines representing water lines along the coast, heavy at first, and becoming lighter and wider apart until they fade into indistinctness, thus .

For coasting, the crayon may be notched and used as we have indicated.

NOTCHED CRAYON.

Many pupils will acquire such skill in map-drawing that they will imitate, and even surpass, in artistic effect, the printed map ; nor are such pupils exceptional cases. Let any teacher persistently and systematically pursue this method, and the results will astonish himself and charm his pupils.

DRAWING COAST-LINES.

II.—HIGHER COURSE OF MAP-DRAWING.

Having for some months diligently put in practice the former method of map-drawing, you are now ready to apply a more exact system.

You will find no royal road to map-drawing by the use of mathematical figures. Each and every continent, island, and state must be known as the painter knows the face he portrays.

Advantages of using Parallels and Meridians rather than Geometrical Figures.—The use of any lines, except those actually impressed on the face of nature, may be objected to as arbitrary; but it is found that the earth's natural marks and boundaries are not sufficient helps in laying off a map. From time immemorial all geographers have agreed to represent the earth as a globe, on which certain *parallels* and *meridians* are drawn. These lines are in universal and daily use by the state-man, the merchant, the seaman, and the explorer. They recur at regular, convenient, and known intervals, and afford all the aid needed in map-drawing. Their points of intersection are as fixed and familiar as the junctions of our great railways; and, although they were originally artificial, they have become *next to natural*, are conceived of as actual furrows in the earth's crust, and are respected as if they were the most ancient landmarks.

I. *To draw a Map after Mercator's Projection* is now the first thing to be done. (See "Mercator's Projection" in Map-Making, p. 160.) In a Mercator's map the meridians of longitude are all parallel. Of course this enlarges the countries toward the poles in their longitude. To preserve the bearing of places on this map, there must be a proportional increase of the degrees of latitude as you go from the equator toward either pole.

Suppose it is desired to draw, at first, an easy map, as that of the State of Kansas. *Having first learned the latitude and longitude of a few places in Kansas, as Fort Dodge, Topeka, etc.*, the scholar is sent to the blackboard, and is directed to proceed according to the following

FORM OF RECITATION.

OUTLINE.—I am to draw the map of Kansas. This State lies between the 37th and 40th parallels of north latitude and the 94th and 102d meridians of west longitude.

As Kansas extends through eight degrees of longitude, draw a line for its northern boundary, and divide it into eight equal parts. The table below shows that one degree of longitude on this boundary (40°) is 46 miles long, against 60 miles for one degree of latitude. Therefore lay off the western boundary of Kansas (which extends through three degrees of latitude) in three parts or degrees, each one being made ⁴⁶/₆₀ or ¾ (nearly) as long as one of the degrees of longitude. Thus I construct the following skeleton for Kansas, approximately accurate.

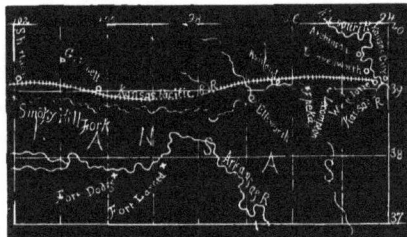

SKELETON FOR KANSAS.

There are mountain chains a little west of Kansas, and the surface of the State is high on the west, and slopes toward the east. The rivers of Kansas, therefore, flow eastward. Smoky Hill Fork comes in on the 39th parallel, and the Arkansas on the 38th, and Red Fork on the 37th parallel. Smoky Hill soon becomes the Kansas, and the Arkansas, on reaching long. 97° 20', flows south. The Missouri cuts off one corner of Kansas. The cities and towns of the State are mostly on its rivers. The railroad generally follows the courses of the rivers.

DIRECTIONS FOR APPLYING THIS METHOD IN PRACTICE.

(1.) In drawing a map of an irregular country, FIRST LAY DOWN PARALLELS AND MERIDIANS SUFFICIENT TO CONTAIN IT. Memorize the latitude and longitude of four or five of the salient points or chief corners of your country, and locate them on the scheme of parallels and meridians. Then draw the outline of the map by connecting these points.

(2.) In drawing *Mercator's Map of the World*, it may be well for the pupil to refer to the last two paragraphs of Lesson XIV., p. 92. Having learned the bearings of the Atlantic coast of North America and South America, the pupil can sketch the opposite shores of the Atlantic by observing how the two great shores of this ocean might fit into each other.

(3.) It will be well, at first, to draw States of easy construction, and afterward the pupil can take more difficult subjects, as Europe.

(4.) Towns should be marked with a round stamp and capital: with a star, thus: ⊚ ✳ .

Railroads and canals may be indicated thus : ▬▬▬▬▬▬▬ RAILROAD.
—— · —— · —— · —— CANAL.

(5.) In drawing, either on blackboard, slate, or paper, the geographical lines should be drawn *lightly*—afterward they may be made heavier, or may be removed, as required.

II. To prepare, with approximate accuracy, a skeleton of converging meridians and curved parallels appropriate for such countries as the United States, North America, Europe, and Asia, all lying wholly one side of the equator, will now be easy of execution.

Take Europe for 'an illustration. OUR UNIT OF MEASURE IS ONE DEGREE OF LATITUDE, OR SIXTY GEOGRAPHICAL MILES.

(1.) By inspecting the map, we find that Europe mostly lies between the 40th and the 70th parallels of N. latitude, and the meridian of Greenwich and the 60th meridian of E. longitude. It therefore extends through 30° of latitude and 60° of longitude.

(2.) Draw a perpendicular line, P (*very light*), cutting the space to be occupied by your map into two equal parts. This is your CENTRAL MERIDIAN.

(3.) As Europe extends through 30 degrees of latitude, lay off on the Central Meridian three equal spaces, each representing 10 degrees of latitude.

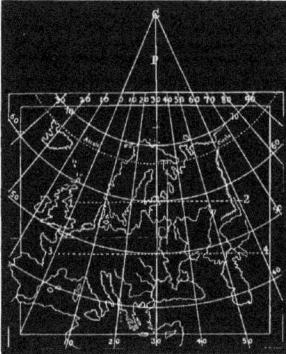

SKELETON FOR EUROPE.

Select those parallels that most nearly divide the latitude embraced by Europe, into three equal parts: these parallels are, in Europe, the 50th and the 60th.

At the 60th parallel (see Table), 1° of longitude is in the ratio of 30 to 60, or ½, of a degree of latitude; the desired 60° of longitude arc, therefore, equal to ½ of 60° of latitude, or equal to the 30° of latitude already marked off on the Central Meridian. Of this length, lay off at the 60th parallel a dotted line 1—2, perpendicular to the Central Meridian and divided by it into two equal parts.

At the 50th parallel (see Table), 1° of longitude is 38. (instead of 30 miles ; therefore, lay off at the 50th parallel the dotted line 3 —4 ³⁸/₆₀, or about ⅘ of that laid off at the 60th parallel.

TABLE OF GEOGRAPHICAL MILES IN A DEGREE OF LONGITUDE AT EVERY FIFTH DEGREE OF LATITUDE FROM THE EQUATOR TO THE POLES.

Lat.	Miles.	Lat.	Miles.	Lat.	Miles.
0°	60	35°	49	65°	25
5°	59	40°	46	70°	20
10°	59	45°	42	75°	15
15°	58	50°	38	80°	10
20°	56	55°	34	85°	5
25°	54	60°	30	90°	0
30°	52				

(4.) This done, draw straight lines through 3 and 1 and 4 and 2, intersecting each other on the C. M. Take the point of their intersection, C, as a centre from which to describe the arcs, or parallels of latitude through the points already marked in the C. M. On the lower parallel divide the distance between the converging lines into six equal parts (10°), and connect these several points with C by straight lines; these lines will represent the meridians.

(5.) To make meridians outside of Europe, you have only to mark, on the 70th parallel, points 10° of longitude apart ; and from C draw straight lines through these points to the margin of the map. Parallels outside of Europe may be made as the others, using C as a centre for describing them. You now have your scheme of parallels and meridians for the map of Europe, after drawing marginal lines to enclose the appropriate space.

(6.) You can now locate the principal capes and indentations on the outline, and *fill in* the whole skeleton as before.

III. To prepare a sketch of parallels and meridians for South America,

Africa, or Australia and the East Indies, lying on both sides of the equator, the parallels are best represented as straight lines and the meridians are traced in curved lines drawn according to the law of the successive decrease in the length of the degrees of longitude.

Draw first the requisite number of parallels and the CENTRAL MERIDIAN, perpendicular to them. Mark off on the equator to the right and left of the C. M. spaces equal to those separating the parallels. In like manner, mark on each parallel corresponding spaces having the same proportion to those on the equator, as a degree of longitude on each respective parallel has to a degree of longitude at the equator, by the table. Curved lines drawn through those successive series of points on each side of the C. M. will represent the meridians. The larger the country, the more inaccurate are methods II. and III.

(IV.) TO DRAW CURVED LINES AND PARALLELS. This may be done with a flexible ruler. The ruler should be made of hickory or ash, $^1/_8 \times ^3/_4$ in.

To use it, set off a few of the points or places through which the meridian or parallel you wish to draw must pass. Then pressing the ends of the flexible ruler against two pins or tacks securely fixed in the drawing-board, bend the ruler with one hand, and draw the curve with the other, as by the following figure.

USING THE FLEXIBLE RULER.

(V.) THE SCALE OF MILES may be made by dividing a degree of latitude into six parts. One part would equal 10 geographical miles; five parts, 50 miles; and ten, 100 miles, etc.

GENERAL HINTS FOR MAP-DRAWING.

(1.) No pupil knows a map until he can draw it from MEMORY.

(2.) THE ORDER FOR DRAWING THE CONTINENTS is, (1.) South America, the least difficult ; (2.) Africa ; (3.) North America ; (4.) Asia ; (5.) Europe.

(3.) It is desirable to give the class a special drill in drawing cities, mountain ranges, hills, peaks, coast-lines, and deserts.

Peaks may be indicated thus :

Deserts are represented by dotted spaces thus :

A DESERT.

(4.) It is often found well, to quicken the attention of the class, to send one pupil to the blackboard with instructions to name and describe, in a clear lively tone, every part of the map as he draws it. But concert recitation should be habitual and always with rapid sketching.

(5.) COPYING ON SLATE OR BLACKBOARD may be done at first with advantage, where the class is composed of beginners. Always, then, begin with small, easily-drawn States, as Colorado, Kansas, Alabama, etc.

(6.) THE MATERIALS necessary for drawing on paper, and coloring, are a lead pencil (No. 2). Bristol-board, a piece of india rubber, pen, cake of India-ink, and gamboge, a plate in which to rub the paints, and one or two camel-hair brushes, and a flat brush for coloring.

(7.) MAPS DRAWN IN LEAD PENCIL should be finished before you commence coloring.

All lines should first be drawn lightly, and afterward corrected. In drawing rivers, begin at the sources, and gradually increase the breadth of your lines as you descend the stream. Draw the backbone of mountains before the spurs.

(8.) Coloring.— Begin by making the boundaries in narrow bright lines, using the camel-hair brush. Do not pass the brush more than once over any part of the map; the brush should be quite full of the tint. To produce a shaded line, take your flat brush and fill one side with water and the other with the tint. Blue is the color for the water.

(9.) India-ink Maps should be first finished in pencil and colored completely; afterward should the mountains, rivers, cities, and shores may be put in with india-ink.

(10.) A Pine Board for Drawing.—The use of a plain soft pine board on which to lay your paper in drawing will be found an important aid.

[Note.—Hitherto your exercises in map-drawing have been practiced with an aim only to approximate accuracy, in impressing the bolder features and general outlines of countries on your mind, and in tutoring the eye and hand to sketch them for the aid of the memory. You may now begin the higher and more satisfactory study of map-making, which is a distinct art in itself.]

MAP-MAKING.

THE first step in map-making is to decide as to the projection. For maps that contain more than 120° of longitude Mercator's is the best. For smaller areas, especially when they are made up chiefly of land, the Rectangular Tangential projection is to be preferred.

I. A MERCATOR'S MAP is the development of the earth's surface on a cylinder, supposed to revolve, tangentially at the equator, upon its own axis once

while you are rolling it on the equator once round the earth. By this development the undulations of longitude are all straight lines, and parallel with each other, and so are the parallels of latitude, but the distance between the latter, as marked on the cylinder, increases as you approach the poles. The advantage of this method is that the course and distance between any two places on the map are straight lines. It is for this reason that all charts used for navigation by every nation are Mercator's.

Suppose we wish to construct a Mercator (say of Kansas) on a scale of one-third of an inch to 60 miles (1°) at the equator. You will then assume 60 meridional parts = 1° = $^1/_3$ inch. Now draw your meridians one-third of an inch apart to represent degrees of longitude. Kansas lies between the parallels of 37° and 40°. There are more meridional parts, as you see by the table, to a degree of latitude between 37° and 40° than there are to a degree near the equator, consequently you must increase the distances between the parallels of Kansas proportionally. Between 37° and 38° lat., there are seventy-five meridional parts = 60, .416 ; between 38° and 39° lat., seventy-seven meridional parts = 60, .427 ; between 39° and 40° lat. there are seventy-eight meridional parts = 60, .433. Mark these distances on the margin of your map sheet ; draw in ink your parallels through them, and you have the skeleton of your Mercator. (See map of Kansas, p. 159.)

Now draw on the map from which you are copying parallels and meridians also for every degree. Thus you have both the original and the copy divided off into sections of 1° square, and you can transfer by the eye and in pencil from one square to the other, first putting in the roads and rivers, as per diagram. This done, fill up with details, then ink, letter, rub out pencil marks, and the angle done. Where great accuracy is required the squares both on the original and the copy should be smaller, so as to contain areas of 30, 20, 10, 5 miles or 1 mile square. A little practice will soon accustom the eye to great accuracy.

The advantage of this plan of working by squares, is that any error that may be made is not carried from one square to another, but is confined within the square to which it belongs. Please look at the Mercator's map, pp. 20, 21, and you will see how the distance between the several parallels of latitude increases as you recede from the equator, and you will moreover see that on maps of this projection alone the north is always at the top, and the east to the right hand, as you have been taught. This rule holds good for no other projection when the map includes a large extent of the earth's surface. For instance, look at the hemispheres, pp. 10 and 11. On the Western Hemisphere the North Pole is to the left of Iceland and to the right on the Eastern. On account of this confusion of bearings of places, especially near the edges of the map, the Mercator projection is generally preferred by physical geographers as well as by navigators for their guidance at sea, and their researches and illustrations.

NUMBER OF MERIDIONAL PARTS IN THE 1st AND EVERY 5th DEGREE OF LATITUDE, FROM THE EQUATOR TO THE 85th DEGREE.

Lat.	Meridional Parts.	Lat.	Meridional Parts.	Lat.	Meridional Parts.
1°	60	30°	69	60°	118
5	60	35°	73	65°	140
10°	61	40°	78	70°	172
15°	62	45°	84	75°	234
20°	64	50°	92	80°	329
25	66	55°	102	85	628

II. THE RECTANGULAR TANGENTIAL PROJECTION is developed by laying together on a flat surface a large number of small planes, nearly rectangular, supposed to have been tangentially placed on a globe and to have received an impression of the country to be mapped. This method is preferred chiefly for the land, as Mercator's is for the sea.

The smaller the country, the more accurate the map ; for on this projection greater accuracy may be developed on a map of America than upon a map of Asia, and greater still upon a map of Europe, which is smaller than either, while on maps of smaller portions of the earth's surface, the room for accuracy is all that can be desired.

This projection has been introduced into chartography by Col. Sir Henry James, of the English Ordnance Survey, who is in charge of the most celebrated map establishment in the world.

The principles of it are very simple, though the mathematical demonstration of them here would be out of place. But after a proper explanation of them and a little thought, you will be able to form a very good idea of them. Imagine, as Sir Henry did, a terrestrial globe of sixty-seven feet, with all places laid down upon it in their true position. Now take a set of plane surfaces, just large enough to cover a space on this globe 4° × 5° in extent. Lay them tangentially on the globe, and side by side over the country to be mapped.

Now suppose, for the sake of illustration, each of these little planes to be transparent, and so prepared as to receive a correct impression of all the geographical features within the 4° × 5° covered by it—somewhat after the manner of the image in the camera of the photographer. Further suppose, that all these little transparent planes, after having received the impression, are taken off the globe and laid side by side on the drawing-board. The theory of the Rectangular Tangential projection may be understood from this illustration. The practical application of it may be learned elsewhere.

www.ingramcontent.com/pod-product-compliance
Lightning Source LLC
Chambersburg PA
CBHW021809190326
41518CB00007B/519

* 9 7 8 3 7 4 1 1 8 7 3 5 3 *